Visual Math DICTIONARY

• Don Balka • Jack Bana
• Colleen Hoover • Linda Marshall • Paul Swan

Copyright © 2007 Didax, Inc., Rowley, MA 01969. All rights reserved.

Reproduction of this book or any unique content from this book by mechanical, electronic, or any other means is hereby prohibited without the express written consent of the publisher.

Printed in the United States of America.

Order number 2-5278
ISBN 978-1-58324-260-5

G H I J K 20 19 18 17 16

395 Main Street
Rowley, MA 01969
www.didax.com

Contents

Foreword

In language it is common for a parent, teacher, or student to make use of a dictionary to check the spelling, meaning, pronunciation, or origin of a word. In mathematics, however, where language can often prove to be a stumbling block, it is rare for a dictionary to be consulted.

In this publication, we have tried to give brief and simple explanations of terms without sacrificing accuracy. When in doubt, we have erred on the side of clarity and simplicity. We tried to put ourselves in the place of the user and therefore kept explanations brief and tried to avoid words that would require further definition. To amplify the meanings, we included many diagrams to illustrate ideas. Where we felt that readers would benefit from an overview of key ideas about a particular topic, we included a detailed reference section in the latter half of the book.

We appreciate that when producing a book of this sort it is difficult to appease the purists while at the same time making it accessible to parents, teachers and students.

List of Mathematical Terms

A

Abacus

A device with beads on wires used to perform calculations.

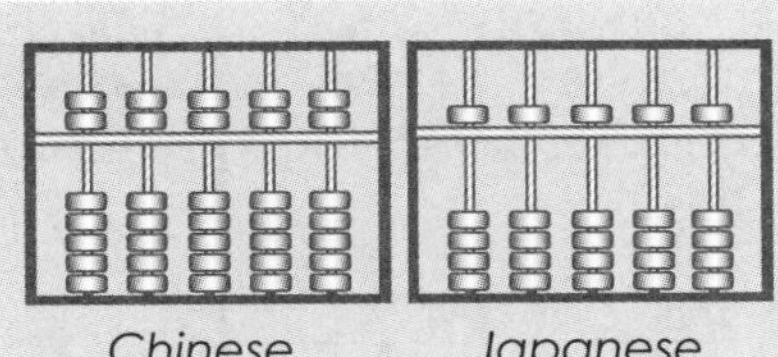

Abcissa

See x-coordinate.

Absolute value (||)

The value of a number no matter the sign. The distance from a number (point) on the number line to 0.

The absolute value of both $^{+}6.8$ and $^{-}6.8$ is 6.8.
$|6.8| = |^{-}6.8| = 6.8$

Acute angle

An angle with a measure greater than 0° and less than 90°.

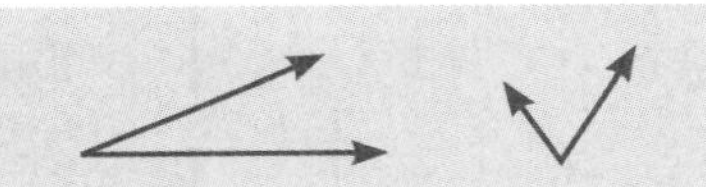

Acute triangle

A triangle in which all three angles are acute.

A.D.

Abbreviation for the Latin *anno Domini* (year of our Lord) applied to years after the birth of Christ. Now often referred to as C.E. (Common Era).

This book was first published in the year 2007 A.D.

Addend

A number being added.

In the sentence 4 + 5 = 9, both 4 and 5 are addends.

Addition (+)

The operation of combining two or more addends to produce another number called a sum.

4 + 3 = 7

Addition property of equality

Adding the same number to each side of an equation produces an equivalent equation.

If $x - 4 = 8$, then $x - 4 + 4 = 8 + 4$. So $x = 12$.

Addition property of zero (Identity property of addition)

When zero is added to any number, the sum is that number [see p. 72].

7 + 0 = 7

Additive identity

Zero is the additive identity for addition of real numbers. When 0 is added to any real number, the sum is that number.

7 + 0 = 0 + 7 = 7
$a + 0 = 0 + a = a$, for any real number.

List of Mathematical Terms

Additive inverse (Opposite)

If the sum of two numbers is zero, then one number is the additive inverse (or opposite) of the other.

4 is the additive inverse of $^-4$ and $^-4$ is the additive inverse of 4 since $4 + (^-4) = 0$.

Adjacent angles

Angles in the same plane with a common ray (side) and a common vertex, such that the interiors do not overlap.

Angles ABC and CBD are adjacent angles. Angles ABC and ABD are ***not*** *adjacent angles.*

Algebra

A generalized form of arithmetic where letters of the alphabet (see variables) are used to represent numbers. [From the Arabic text Al-jabrwa'l muqabalah (ca 825 A.D.).]

$$3x + 2y = 6$$
$$x^2 + 2x + 1 = 10$$

Algebraic expression

An expression involving numbers, variables and/or operations. An algebraic expression is **not** an equation.

$2x + 5y$; $3x^2$; $a + 7$

Algorithm

A step-by-step method or procedure for solving a problem.

Alternate exterior angles

The exterior angles on opposite sides of a transversal that cuts parallel lines.

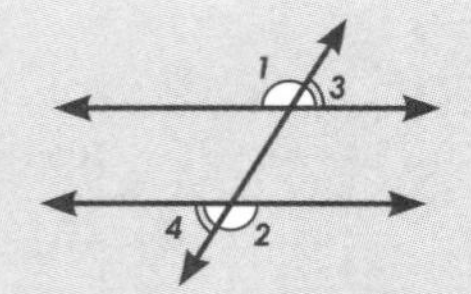

∠1 and ∠2 are alternate exterior angles. ∠3 and ∠4 are alternate exterior angles.

Alternate interior angles

The interior angles on opposite sides of a transversal that cuts parallel lines.

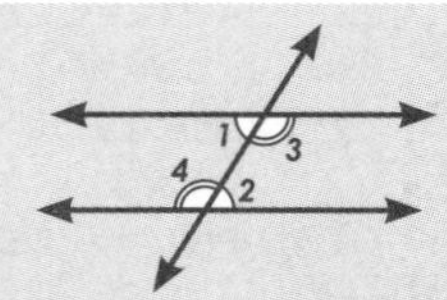

∠1 and ∠2 are alternate interior angles. ∠3 and ∠4 are alternate interior angles.

Altitude

A height or perpendicular distance.

The helicopter was flying at an altitude of 300 meters.

List of Mathematical Terms

Altitude (of a triangle)

A line segment from a vertex, perpendicular to the opposite side (or the opposite side extended).

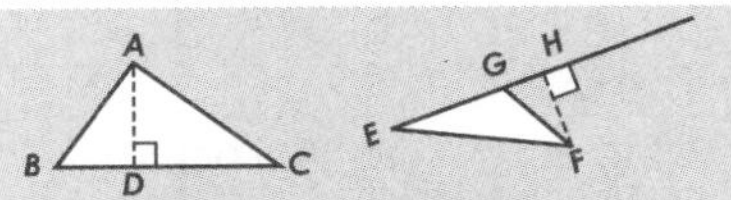

$\overline{AD}$ is an altitude.

$\overline{FH}$ is an altitude. It is on the exterior of the triangle.

a.m. (am)

The time between midnight (12:00 a.m.) and noon (12:00 p.m.); Latin abbreviation for ante meridiem.

We ate breakfast at 7:30 a.m.

Angle

The union of two rays with a common endpoint, called the vertex.

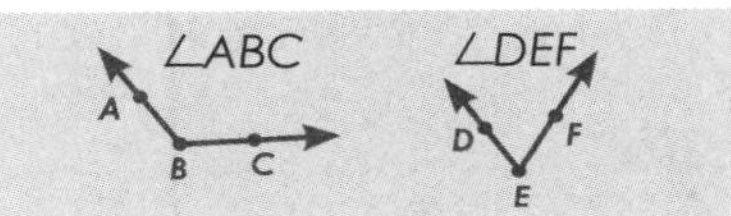

Annual

Occuring once each year.

Annual interest rate

The percent of the principal earned each year.

If $100 is deposited in a bank account that pays 5% per year, then the annual interest rate is 5%.

Annulus

The plane region between two concentric circles.

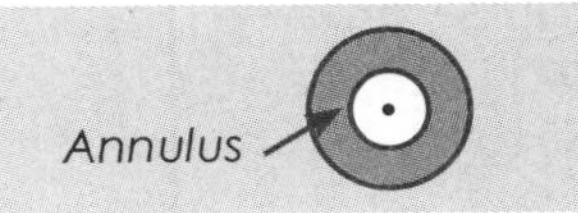

Apex

In a pyramid or cone, the vertex opposite the base.

The top point of a pyramid.

Approximation

An estimated or inexact value.

78 x 99 ≈ 7800, since 78 x 100 = 7800

Arc ($\widehat{AB}$)

1. Part of a circle that can be drawn without lifting a pencil.
2. A connection between two vertices in a network.

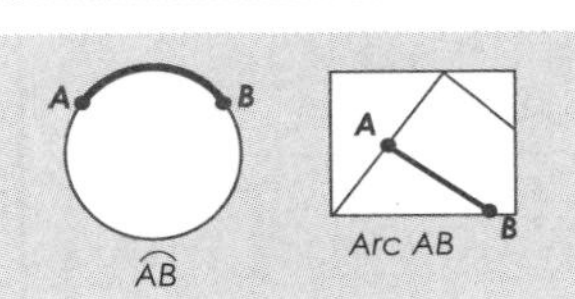

Are

A metric unit of are equivalent to $100m^2$.

Area

The amount of surface.

The area of this page is about 400 square centimeters.

List of Mathematical Terms

Arithmetic

The part of mathematics involving the study of numbers and their operations and properties.

Arithmetic mean

The sum of the values in a data set divided by the number of values.

The arithmetic mean of 4, 7, and 13 is (4 + 7 + 13) ÷ 3 = 8.

Arithmetic sequence (progression)

A sequence of numbers where the next term is generated by adding or subtracting a constant value from the previous term.

The sequence 4, 7,10, 13, 16 ... is an arithmetic sequence. So is 17, 12, 7, 2, $^{-}3$, $^{-}8$ ...

Array

An arrangement of objects into rows and columns.

The 12 stars shown are in an array of 3 rows and 4 columns. ★★★★ ★★★★ ★★★★

Ascending order

An ordered arrangement according to number or size, beginning with the smallest.

6, 10, 14, 18 are in ascending order.

Associative property of addition

When adding three or more numbers, the grouping does not affect the sum [see p. 72].

2 + (3 + 4) = (2 + 3) + 4

Associative property of multiplication

When multiplying three or more numbers, the order in which they are multiplied does not affect the product [see p. 72].

3 x (5 x 6) = (3 x 5) x 6

Asymmetry (asymmetrical)

Where an object or figure has no reflection symmetry. See also Symmetry p. 113.

The figures shown are asymmetrical. **N**

Attribute

A trait or characteristic.

One attribute of a quadrilateral is that it has four sides.

Attribute blocks

A set of plastic or wooden blocks consisting of differing attributes of shape, color, size and thickness such that no two blocks in the set are the same.

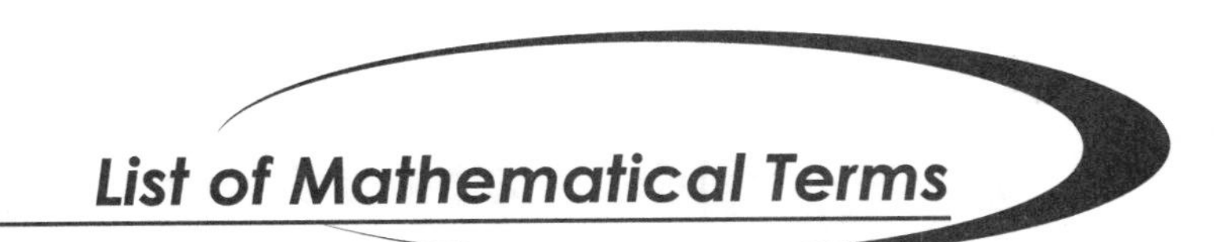

List of Mathematical Terms

Average

A single number used to describe what is typical of a set of data. The (arithmetic) mean, median and mode are examples of averages.

Axis (axes)

A linear direction, usually vertical or horizontal.

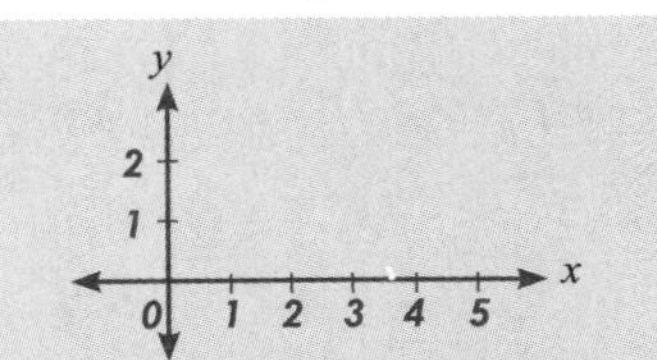

A bar or column graph and the coordinate plane each have both vertical and horizontal axes.

Axis of symmetry

See Line of symmetry.

B

Balance

1. Equipment using a pivoted beam to compare the masses of objects, or to weigh objects.
2. The amount of money in an account.

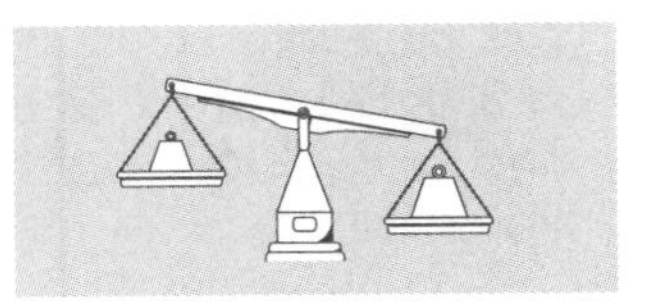

Bar graph

A graph in which the lengths of the bars are used to represent and compare data.

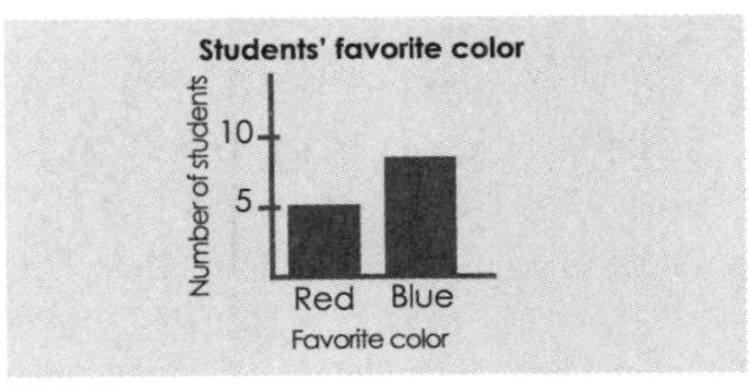

Base (of a cone)

The circular face.

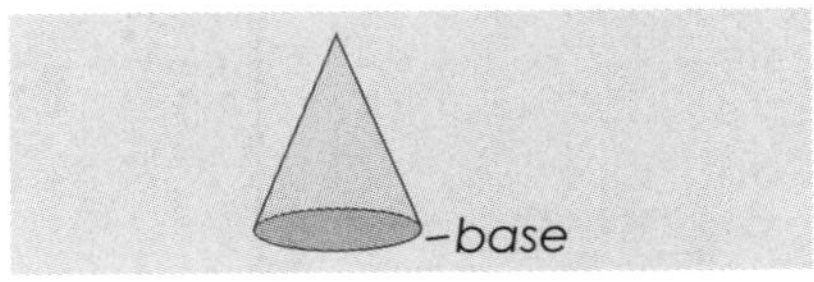

Base (of a place value number system)

In a place value numeration system, the grouping that is used. The decimal numeration system is a base 10 numeration system.

Base (of a power)

The repeated factor in a power.

In the power 4^3, 4 is the base. In the power $(x + 2)^5$, $x + 2$ is the base.

List of Mathematical Terms

Base (of a pyramid)

The polygonal face opposite the apex.

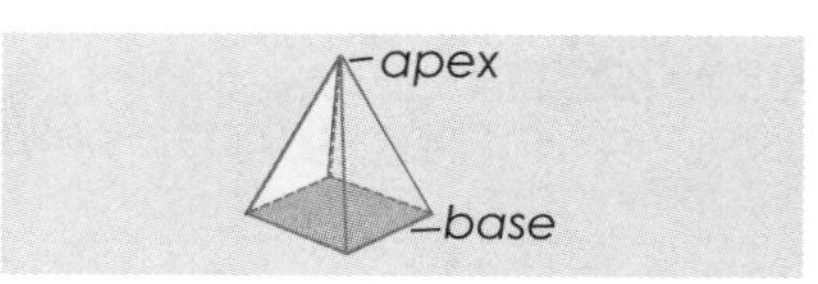

Bases (of a prism or cylinder)

The faces that are congruent and parallel.

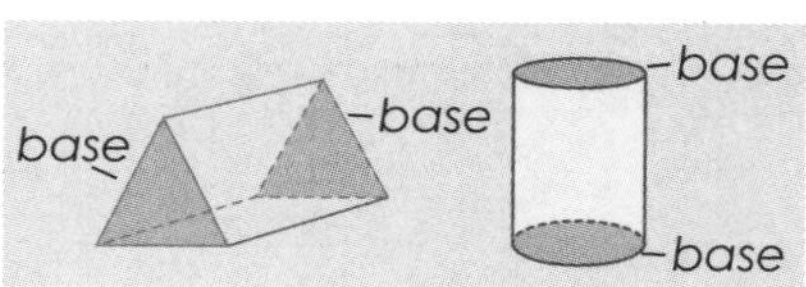

Bases (of a trapezoid)

The parallel sides of a trapezoid.

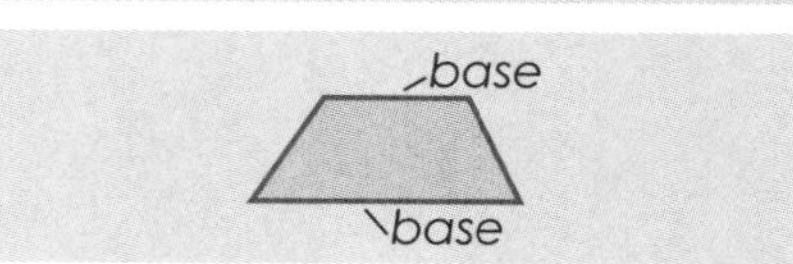

Base ten or decimal (numeration) system

A system of numeration in which the place values are powers of ten.

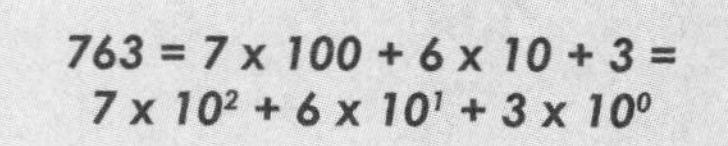

Base ten blocks

Wooden, plastic, or foam materials, consisting of four different pieces called units, longs, flats and blocks, to represent place values in the decimal system of numeration.

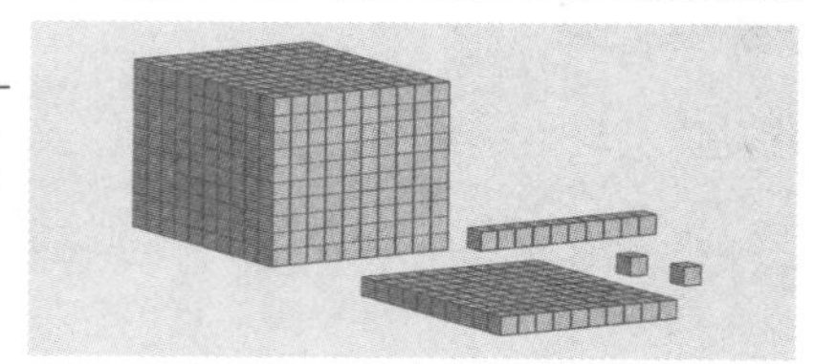

Basic (number) facts

The number facts needed to be able to carry out all calculations; i.e. the addition facts from 0 + 0 to 9 + 9 and the related subtraction facts, together with the multiplication facts from 0 x 0 to 9 x 9 and the related division facts [see pp. 70–71]. There are 100 basic addition facts, 100 basic subtraction facts, 100 basic multiplication facts and 90 basic division facts.

B.C.

Abbreviation for Before Christ and applied to dates. Now often referred to as B.C.E., before the common era.

The Roman emperor Julius Caesar was born in the year 100 B.C.

Biased sample

A sample that is not representative of the population from which it is selected.

The heights of professional basketball players form a biased sample of the heights of U.S. citizens.

Bicentennial

Two-hundredth anniversary.

1988 was the bicentennial of the first European settlement in Australia.

Billion
The number 1,000,000,000 or 10^9 [see p. 77].

Binary
Involving two possibilities such as yes/no, true/false, 0/1, etc.

Binary numbers
Numbers in the base two numeration system, which uses 0s and 1s. They are the basis of all digital systems.

The binary number 11011 represents
$$1 \times 2^4 + 1 \times 2^3 + 0 \times 2^2 + 1 \times 2^1 + 1 \times 2^0 = 27$$
in base 10.

Binomial
A polynomial with two terms.

$3x + 4; x^2 - 2x$

Bisect
To divide into two congruent parts.

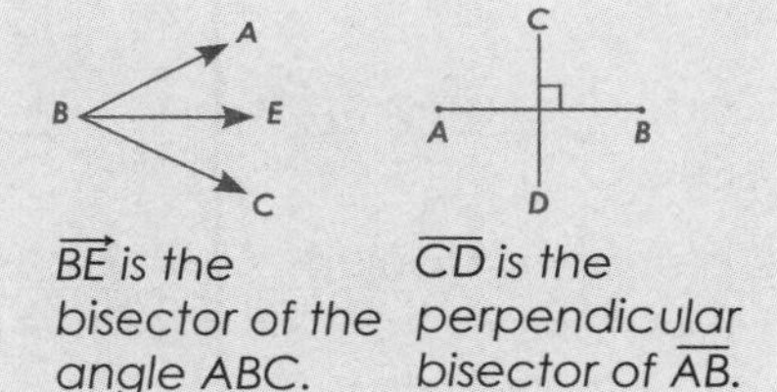

$\overrightarrow{BE}$ is the bisector of the angle ABC.
$\overline{CD}$ is the perpendicular bisector of $\overline{AB}$.

Box and whisker plot (Box plot)
A data display that divides a data set into four parts using the lower extreme value, lower quartile value, median, upper quartile value, and extreme value.

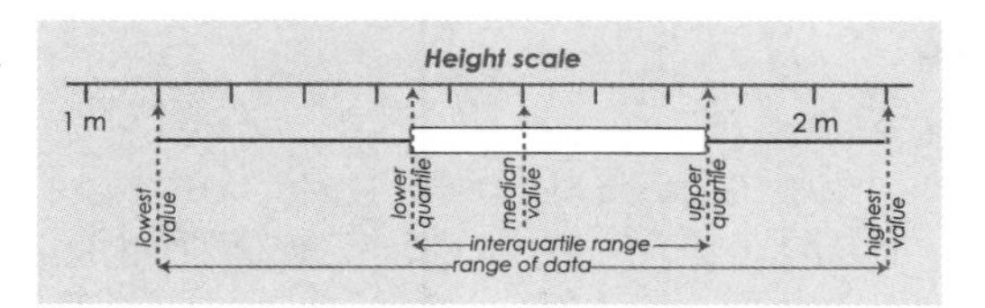

Breadth (Width)
The distance across.

The breadth of this page is about 17 cm.

Calculate
Use a mathematical procedure to determine a number, quantity, or expression.

Calculator
An electronic calculating device.

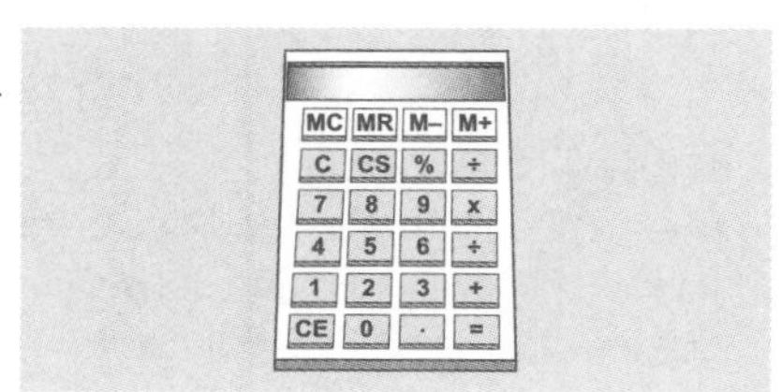

Calendar
A chart showing the days, weeks and months of the year [see pp. 121–122].

List of Mathematical Terms

Capacity
The amount a container can hold.

The capacity of the bottle is one liter.

Cardinal number
The number of objects in a set. The answer to the question "How many?"

For the set {a, b, c, d} the cardinal number is 4.

Carroll diagram
A grid-like structure for categorizing results.

Colored blocks sorted into a 2 x 2 Carroll diagram.

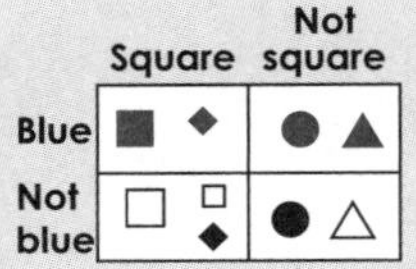

Cartesian coordinates
See Coordinates.

Cartesian product (A x B)
All possible matchings of the members of one set with the members of another set, illustrating multiplication.

For sets A= {a, b, c} and B = {1, 2} the Cartesian product is the set {(a, 1), (a, 2), (b, 1), (b, 2,) (c, 1), (c, 1)}.

Celsius (C)
The basic metric unit for temperature measurement. The freezing point of water is 0 degrees and the boiling point is 100 degrees. [Named after its 18th century inventor, Anders Celsius. See p. 120.]

Cent (¢)
A currency or money unit of one-hundredth of a dollar.

Centi
A basic metric prefix for $\frac{1}{100}$ of a whole unit.

Commonly used units are centimeter (cm), centiliter (cL), and centigram (cg).

Centimeter (cm)
A unit of length in the metric system which is $\frac{1}{100}$ of a meter.

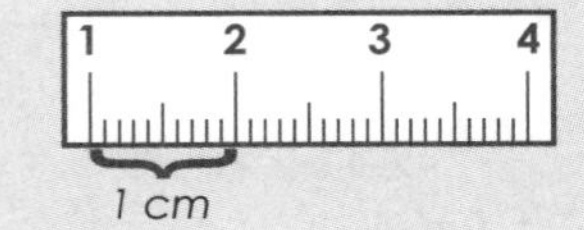

Center (of a circle)
The point in the interior of a circle that is the same distance from all the points on the circle.

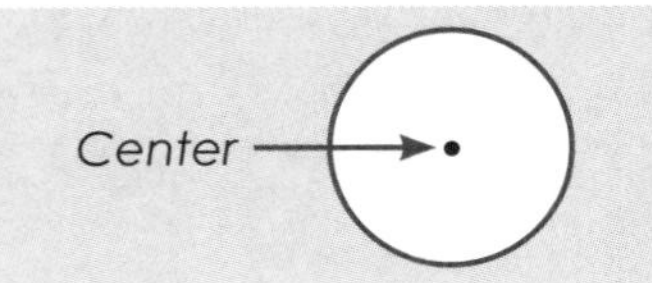

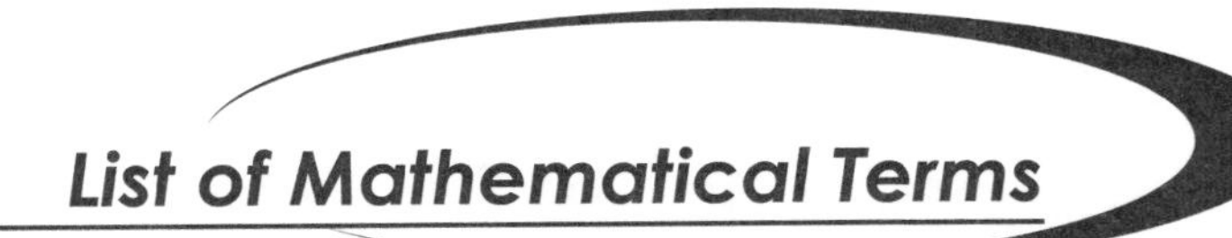

List of Mathematical Terms

Center (of a sphere)
The point in the interior of a sphere that is the same distance from all points on the sphere.

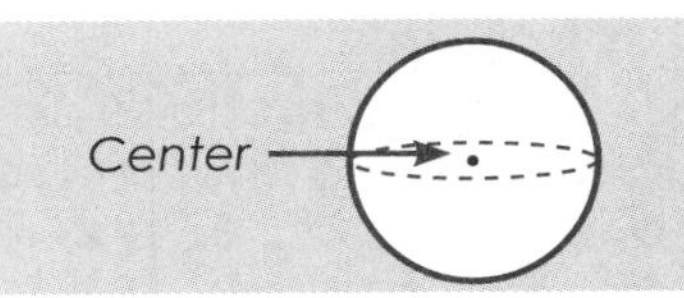

Century
Usually referring to 100 years.

This is the 21st century A.D.

Center (of rotation)
The point about which a figure is turned when the figure undergoes a rotation.

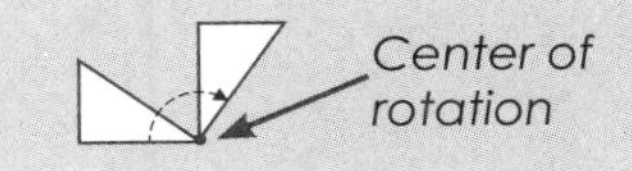

Certain (event)
An event that will happen for sure. An event with a probability of 1.

In tossing a dime, the event "heads or tails" is certain.

Chance (Probability)
The likelihood or probability of an event occurring.

In tossing a coin, the chance (probability) of getting "heads" is $\frac{1}{2}$.

Chord
A line segment joining two points of a circle [see p. 106].

Line segment $\overline{AB}$ is a chord. Diameter $\overline{CD}$ is a chord.

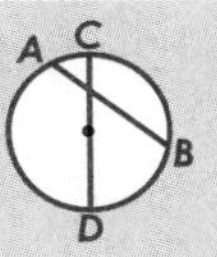

Circle
The set of all points in a plane that are the same distance from a fixed point, called the center. One of the conic sections.

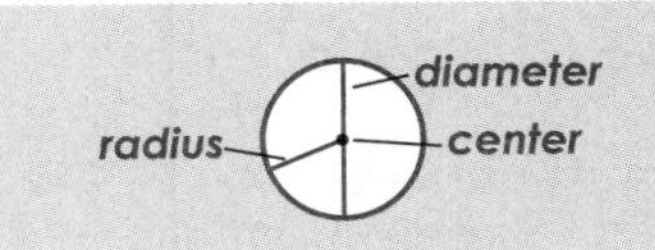

Circle graph (Pie chart, Pie graph)
A graph that represents data as parts of a circle. The entire circle represents all of the data.

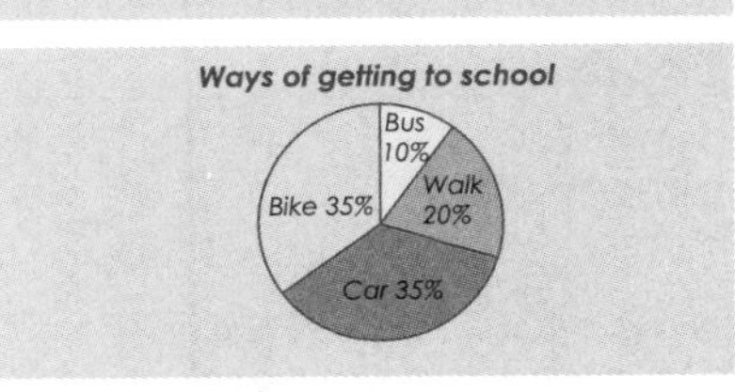

Circumference
The distance around a circle. The perimeter of a circle [see p. 106].

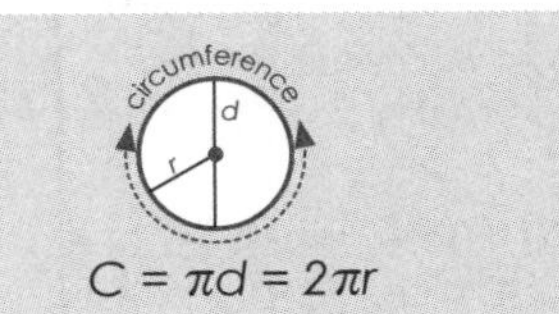

$C = \pi d = 2\pi r$

Class boundary
The border between two class intervals.

The boundary between the intervals 31–35 and 36–40 is 35.5.

List of Mathematical Terms

Class interval

A category of grouped data described by an interval. The endpoints of the interval are called class limits. For some categories there are no limits.

class intervals

Age	Frequency
1–10	4
11–20	7
21–30	3
Over 30	2

Clockwise

In the same direction as the movement of the hands of the clock.

Closed (plane) curve

A curve that begins and ends at the same point.

All five shown are simple closed curves.

Clustering

A method of estimating a sum when numbers being added have about the same value.

An estimate of the sum of 63 + 58 + 55 is about 3(60), since the 3 values cluster around 60.

Coefficient

The number part of a term that includes a variable.

In the expression $3x + 4y$, 3 is the coefficient of x and 4 is the coefficient of y.

Collinear

Three or more points in the same line.

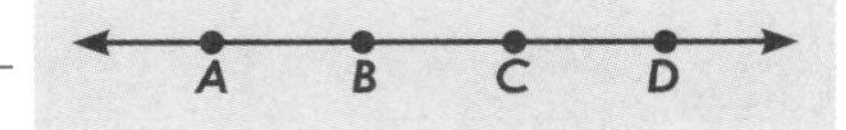

Column

A vertical arrangement of objects.

Combination

A grouping of objects in which the order is not important.

In selecting 2 letters from the word CUBE, there are 6 combinations: CU, CB, CE, UB, UE, BE. Note that CB is the same grouping as BC.

Common denominator

A common multiple of the denominators (bottom numbers) of two or more fractions.

Common denominators of the fractions $\frac{1}{2}$ and $\frac{1}{3}$ are 6, 12, 18, 24, 30, …

Common factor

A whole number that is a factor of two or more nonzero whole numbers.

The common factors of 8 and 12 are 1, 2 and 4.

Common multiple

A whole number that is a multiple of two or more nonzero whole numbers.

The common multiples of 3 and 4 are 0, 12, 24, 36, 48, …

Commutative property of addition

When adding two numbers, the order does not affect the sum [see p. 72].

7 + 3 = 3 + 7; $a + b = b + a$

Commutative property of multiplication

When multiplying two numbers, the order does not affect the product [see p. 72].

5 x 2 = 2 x 5; $a \times b = b \times a$

Compass

1. An instrument used for drawing circles.
2. An instrument for finding directions. It has a magnetized needle that always points to the north.

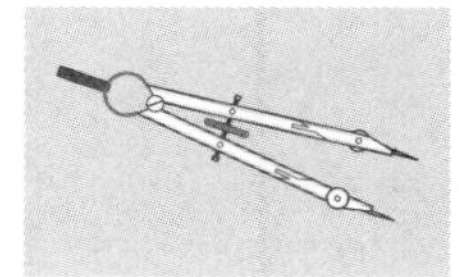

Compatible numbers

Numbers that make a calculation easier.

In adding 8 + 6 + 5 + 2 + 4, 8 and 2 are compatible numbers, since their sum is 10; likewise, 6 and 4 are compatible numbers.

Complementary angles

Two angles whose measures have a sum of 90˚ [see p. 98].

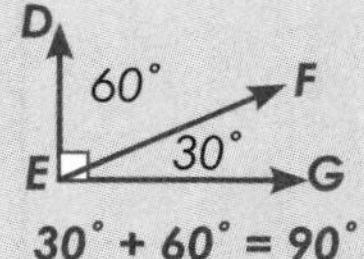

30˚ + 60˚ = 90˚

∠DEF and ∠FEG are complementary angles.

Complementary events

Events that have no outcomes in common and that together contain all the outcomes of the experiment.

In rolling a number cube, the events "getting an even number" and "getting an odd number" are complementary events.

Composite number

A whole number greater than 1 that has factors other than 1 and itself.

6 is a composite number since its factors are 1, 2, 3 and 6.

Concave

A shape is concave if there is at least one line segment connecting two interior points which passes through the exterior of the figure.

Both shapes are concave.

Concentric circles

Circles having the same center.

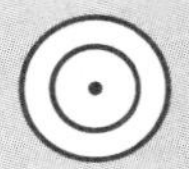

List of Mathematical Terms

Cone
A solid that has one circular base, a vertex (apex) that is not in the same plane as the base, and a curved surface connecting the two.

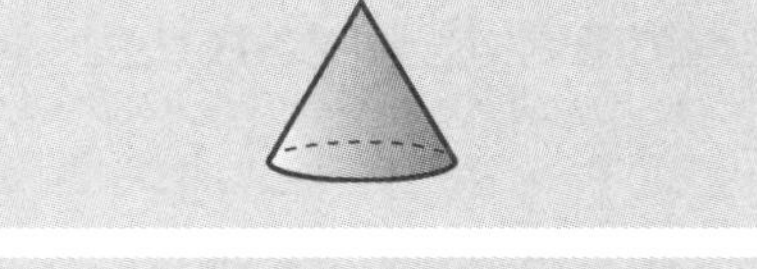

Congruent angles (≅)
Angles that have the same measure.

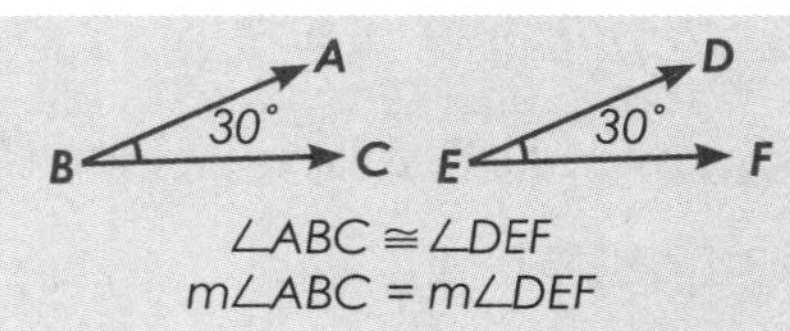

Congruent figures
Figures having the same size and shape.

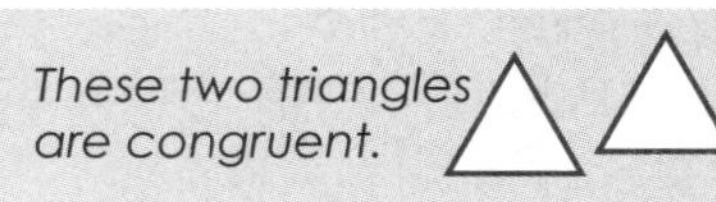
These two triangles are congruent.

Conic sections
The plane figures that result when a cone is cut by a plane: circle, ellipse, parabola, hyperbola.

The conic section shown here is a circle.

The conic section shown here is an ellipse.

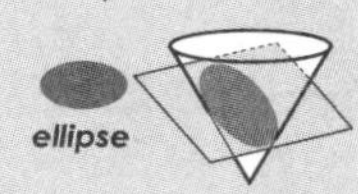

Consecutive numbers
Numbers that follow each other in a sequence.

5, 6, 7, 8, 9 and 0.1, 0.2, 0.3, 0.4, 0.5 are both sets of consecutive numbers.

Constant (term)
1. A term that has a number but no variable.
2. A fixed quantity.

In the expression $3x + 7$, 7 is the constant term and $3x$ is the variable term.
The speed of light is constant.

Continuous data
Data that can take on any value in an interval.

Temperatures, weights and heights are examples of continuous data.

Contraction (Reduction)
A transformation that reduces the size of a figure.

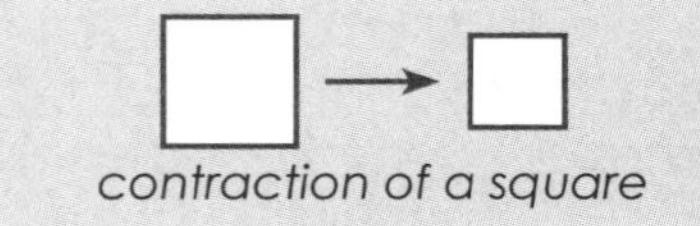
contraction of a square

Convex
A shape is convex if every line segment connecting two interior points lies entirely on the interior points lies entirely on the interior of the figure.

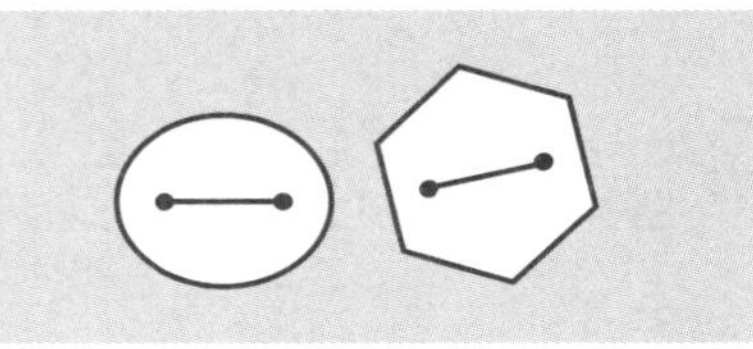

Coordinate plane (Coordinate grid)

A plane divided into four quadrants by a horizontal number line, typically called the x-axis, and a vertical number line, typically called the y-axis.

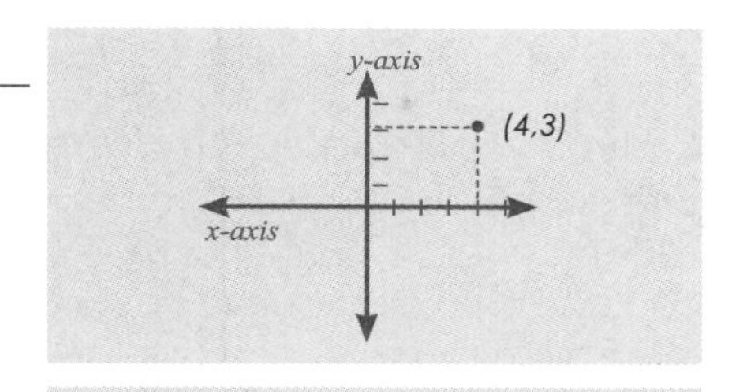

Coordinates

The numbers in an ordered pair that locate a point on a coordinate plane.

In the above illustration, the ordered pair (4,3) represents a point 4 units to the right of (0,0) and 3 units up from (0,0).

Coplanar

In the same plane.

Corresponding angles

1. Angles that occupy corresponding positions when a line intersects two parallel lines.
2. Angles in like or similar positions in different figures, particularly congruent figures.

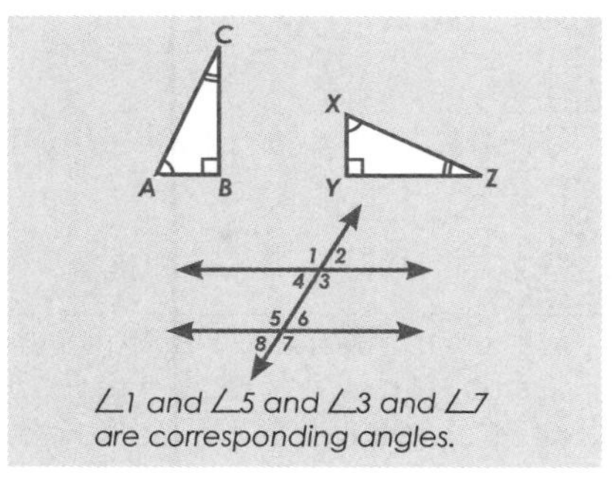

∠1 and ∠5 and ∠3 and ∠7 are corresponding angles.

Corresponding sides

Sides in like positions in different polygons, particularly congruent or similar polygons.

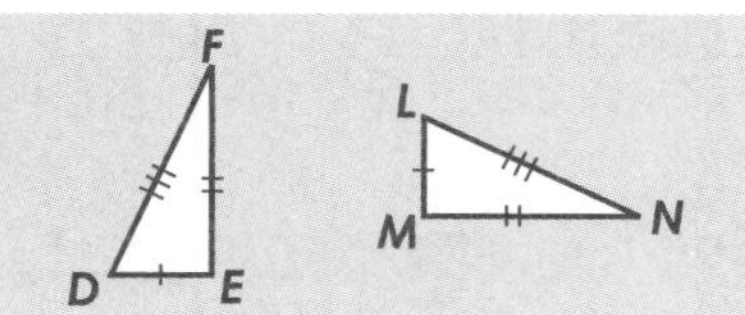

Cosine

The cosine of any acute angle in a right triangle is the ratio of the length of the adjacent leg to the length of the hypotenuse.

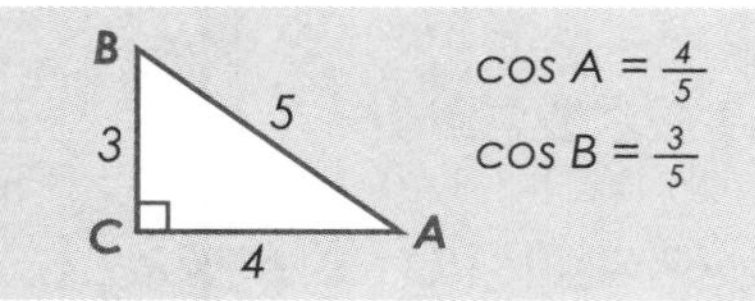

Counter clockwise (Counterclockwise)

Circular direction about a clock face in the opposite direction of the movement of the hands of the clock.

Counterexample

An example that contradicts a statement.

Statement: *All primes are odd.*
Counterexample: *2 is prime and is even.*

List of Mathematical Terms

Counting numbers (Natural numbers)

N = {1, 2, 3, 4, 5, ...}

Counting principle (Fundamental counting principle)

If one event can occur *m* ways and a second mutually exclusive event can occur *n* ways, then the number of ways the two events can occur together is *m* x *n*. The counting principle can be extended to more that two events.

Tossing a coin has 2 outcomes. Tossing a 6-sided number cube has 6 outcomes. Together, the experiment has 2 x 6 = 12 outcomes.

Cross product (Cross multiplication)

A term used in solving a proportion. In a proportion *a*/*b* = *c*/*d*, the products *ad* and *bc* are cross products.

$\frac{2}{3} = \frac{4}{6}$, *so 2(6) = 3(4).*

Cross-section

The plane figure resulting from a solid cut by a plane.

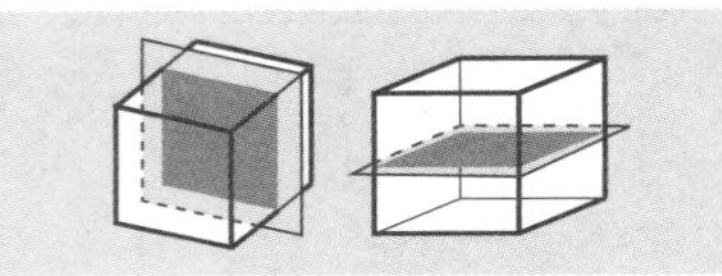

Cube

A rectangular prism with six congruent square faces. A cube is also a regular hexahedron, one of the five Platonic solids.

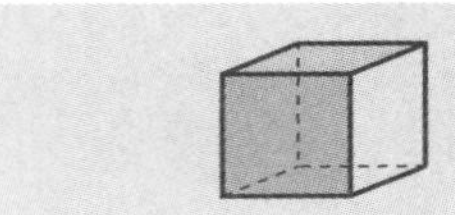

Cube root (Cubic root)

The base of a cubic number.

The cube root of 64 is 4. The cube root of 125 is 5. The cube root of 9^3 is 9.

Cubic centimeter (cm^3)

A unit of volume occupied by the equivalent of a cube 1 cm x 1 cm x 1 cm in size.

The pictured cube is approximately 1 cm^3.

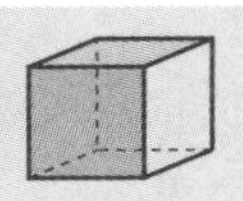

Cubic meter (m^3)

A unit of volume occupied by the equivalent of a cube 1 m x 1 m x 1 m in size.

Cubic number (Cube)

A number obtained by multiplying another counting number by itself twice. A number raised to the third power.

1, 8, 27, 64 and 125 are cubic numbers.
$1 = 1 \times 1 \times 1 = 1^3$, $8 = 2 \times 2 \times 2 = 2^3$, $27 = 3 \times 3 \times 3 = 3^3$, $64 = 4 \times 4 \times 4 = 4^3$

Cuboid

A rectangular prism that is not a cube.

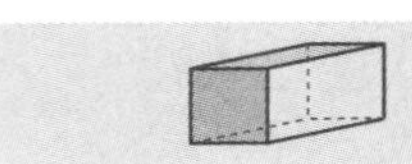

Cumulative frequency

In a frequency distribution with intervals, the total frequency of all values less than or equal to a denoted interval.

Number of children	Number of couples (f)	Cumulative frequency (cf)
0	4	4
1	7	11
2	24	35

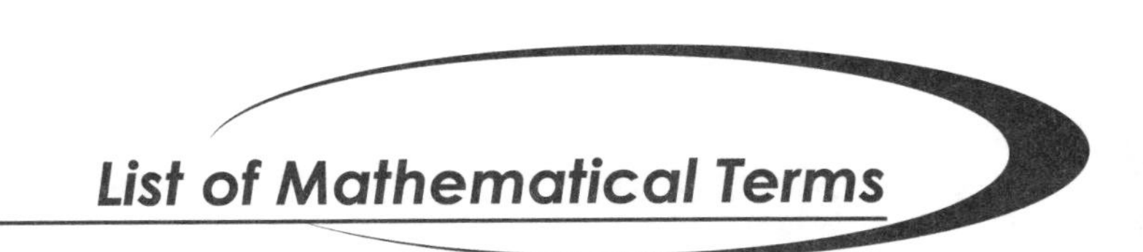

Curve
A path that can be drawn in a plane without lifting a pencil.

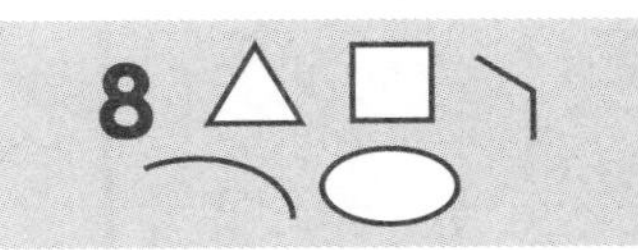

Cylinder
A solid formed by two congruent circular bases that lie in parallel planes, joined by a curved surface.

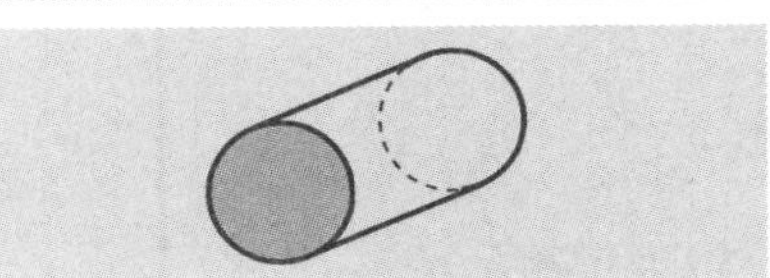

D

Data
Information, facts, or numbers that describe something. Data is plural; datum is singular.

The number of cars, trucks, and motorcycles entering an intersection in a 1-hour time block.

Decade
A 10-year period of time.

This book was published in the first decade of the 21st century.

Decagon
A 10-sided polygon [see p. 99].

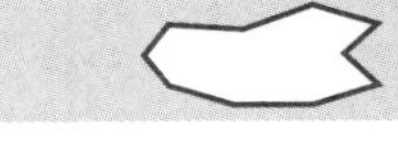

Decahedron
A 10-faced polyhedron.

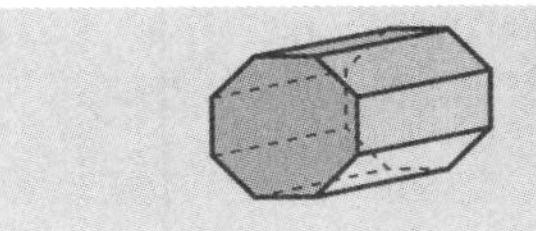

Decimal
A number that is written using the base ten numeration system. Each place value is ten times the place value to the right.

0.7 24.75 256.1 3.5

Decimal fraction
A fraction whose denominator is a power of 10.

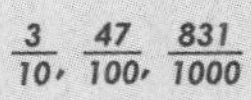

Decimal point (.)
The symbol between the ones and the tenths in a decimal.

4.6 is read as "4 and 6 tenths." 9.75 is read as "9 and 75 hundredths."

Degree (°)
1. A unit of measure of an angle. A complete rotation about a circle has a measure of 360°. 1° is $\frac{1}{360}$ of a circle.
2. A unit of measure of temperature.

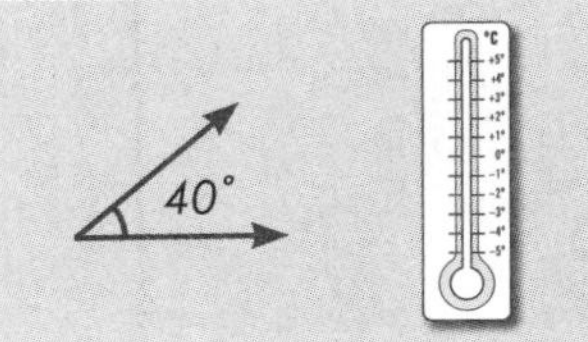

List of Mathematical Terms

Denominator
The nonzero number b in the fraction $\frac{a}{b}$.

The denominator of $\frac{5}{12}$ is 12.

Density
The mass per unit volume of material.

The density of water is set at 1 because 1 cm^3 of water has a mass of 1 g, but the density of iron is 7.87 since 1 cm^3 of iron has a mass of 7.87 g.

Dependent events
Two events such that the occurence of one affects the likelihood that the other will occur.

A bag contains 3 red balls and 5 white balls. Two balls are randomly drawn, without replacement. The event "First ball is red" and "Second ball is red" are dependent events.

Depth
Vertical distance downwards from a given position. See Altitude.

Descending order
An ordered arrangement according to number or size, beginning with the largest.

16, 13, 10, 7 are in descending order.

Diagonal
A line segment joining two vertices that are not next to each other.

In the rectangle ABCD, the diagonals are the line segments AC and BD.

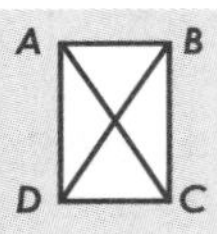

Diameter (of a circle)
A chord that passes through the center of a circle [see p. 106].

The diameter is $\overline{AB}$.

Die (Dice)
See Number cube.

Difference
The result when one number is subtracted from another.

The difference between 10 and 6 is 4.

Digits
1. A symbol used in a place value system.
2. A digit is also a finger. [From the Latin *digitus*.]

0, 1, 2, 3, 4, 5, 6, 7, 8, 9 are the digits used in our Hindu-Arabic decimal system.

In a binary system, the digits are 0 and 1.

Dilation
A transformation that expands (stretches) or contracts (shrinks) a figure. See Size transformation.

Dimension
The extent of a figure, particularly the length (l), width (w), and height (h). A line has a dimension of 1; a plane has a dimension of 2. a solid figure has a dimension of 3.

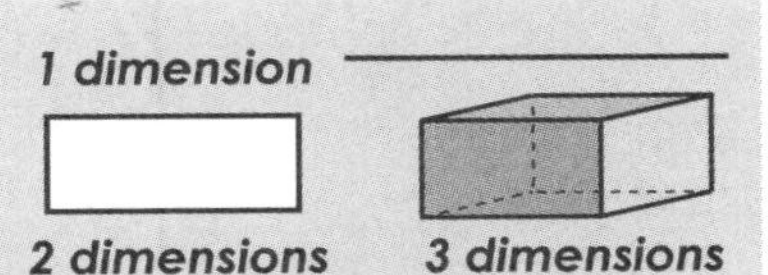

Directed numbers
Numbers that are left (negative) or right (positive) of 0 on a number line.

Direct proportion
Two quantities are in direct proportion when both increase or decrease at the same rate.

Chocolate bars	Cost
1	$2.50
2	$5.00
3	$7.50

Discount
The amount by which the original price of an item is reduced.

If an item is marked at $50 and sold for $40, there has been a discount of $10 or 20%.

Discrete data
Information that can be recorded with only a finite or countable set of numbers.

The costs of various candy bars are 70¢, 75¢, 79¢, 80¢, 85¢ and $1.20.

Disjoint events
Events that have no outcomes in common.

When rolling a number cube, the events "getting an even number" and "getting an odd number" are disjoint events.

Disjoint sets
Sets that have no common members.

The set of odd numbers and the set of even numbers are disjoint sets.

Distance
The length of a path—real or imagined—between two points or objects.

The distance across the road is 10 m, while the distance from the South Pole to the Equator is about 10,000 km.

Distance formula
For two points (x_1, y_1) and (x_2, y_2) in the coordinate plane, the distance between them is given by the formula $d = \sqrt{(x_1 - x_2)^2 + (y_1 - y_2)^2}$.

The distance between (6, 8) and (3, 4) is

$$d = \sqrt{(6-3)^2 + (8-4)^2} = \sqrt{9 + 16} = \sqrt{25} = 5$$

Distortion
A transformation that changes the shape of a figure.

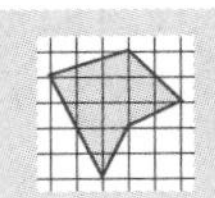
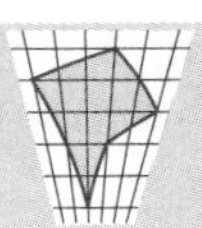

List of Mathematical Terms

Distributive property (of multiplication over addition)

A number and a sum can be multiplied by multiplying the number by each addend and then adding these products. The same property applies to subtraction [see p. 72].

$$a\,(b + c) = ab + ac$$
$$a\,(b - c) = ab - ac$$

Dividend

The number that is divided by another number.

In the expression 35 ÷ 7, the dividend is 35.

Divisible

An integer is divisible by a nonzero integer when the remainder is 0.

A number is divisible by a second number if the second number is a factor of the first.

35 is divisible by 7.

Divisor

The number to be divided into another number.

In the expression 35 ÷ 7, the divisor is 7.

Division (÷)

The operation of dividing one number (the dividend) by another number (the divisor) to obtain a third number (the quotient). See the three items above.

Dodecagon

A polygon with 12 sides [see p. 99].

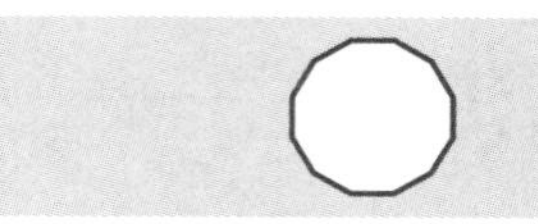

Dodecahedron

A polyhedron with 12 faces. A regular dodecahedron is one of the Platonic solids.

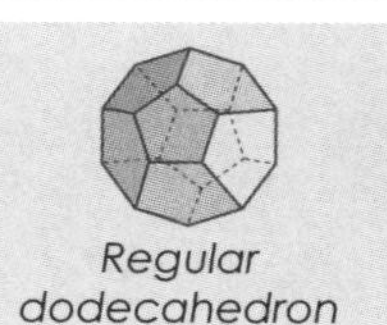

Regular dodecahedron

Decagonal prism

Dollar

A unit of currency consisting of 100 units normally called cents.

Domain (of a function)

The set of all input values for a function.

Input	1	2	3	4
Output	3	6	9	12

Domain = {1, 2, 3, 4}

Dot plot

A data display consisting of a horizontal number line on which each data point is denoted by a dot or an X above the corresponding number line value.

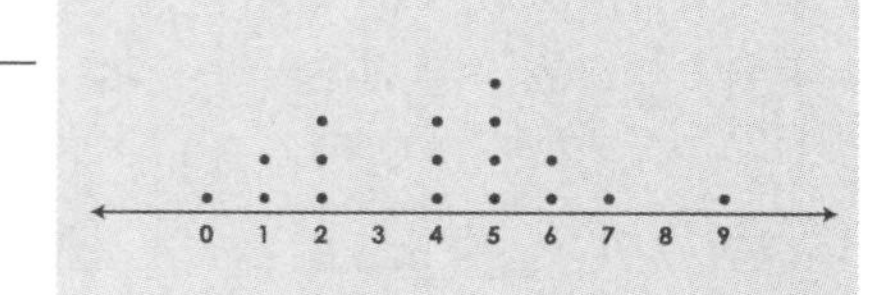

Double

Twice the number or amount.

Double 8 is 16.

Double bar graph

A bar graph that shows two sets of data on the same graph.

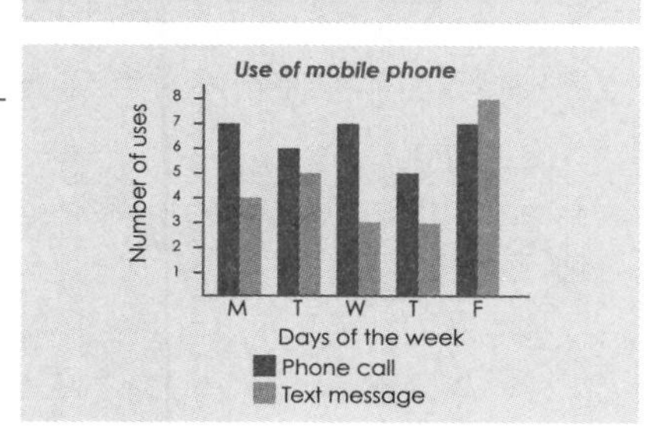

Dozen

Another name for 12.

E

Edge

A line segment where two faces of a polyhedron meet.

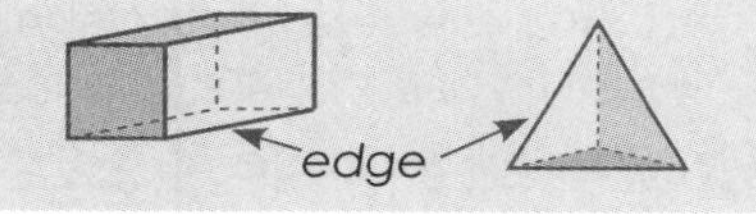

Elapsed time

The amount of time between a start time and an end time.

The elapsed time between 8:00 a.m. and 11:45 a.m. is 3 hours and 45 minutes.

Element ($\in$)

The member of a set.

In a set A = {1, 2, 3, 4, 5}, 4 is an element. We write $4 \in A$.

Ellipse

A closed curve that results when a cone is cut by a plane. One of the conic sections.

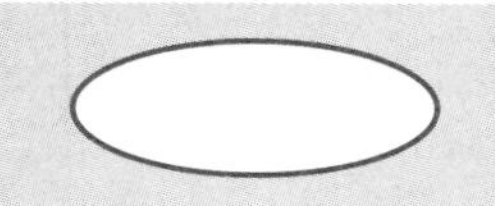

Empty set ({ } = Ø)

A set with no members.

The set of whole numbers greater than 1 and less than 2.

Endpoints

The boundary values of an interval.

For the interval 20–29, the numbers 20 and 29 are endpoints.

English system (of measurement)

A system of measurement still prevalent in the U.S., often referred to as the Imperial system; predates the metric system.

Enlargement

See Expansion [see p. 115].

Equal sets

Sets containing the same elements.

Set A = {1,2,3} and Set B = {2,3,1} are equal sets: A = B.

List of Mathematical Terms

Equal sign (=)
A math symbol that indicates two expressions name the same number or quantity.

6 + 4 = 3 + 7 may be read as "six plus four equals three plus seven."

Equally likely events
Events that have the same probability of occurring.

In tossing a coin, the event "getting a head" and the event "getting a tail" are equally likely events since the probability of each occurring is $\frac{1}{2}$.

Equation
A mathematical sentence stating that two expressions are the same.

10 + 9 = 19
4 + 7 = 3 + 2 + 6
$3y$ + 1 = 28

Equiangular triangle
A triangle with three congruent angles. All equiangular triangles are equilateral triangles.

Equilateral triangle
A triangle with three sides of the same length.

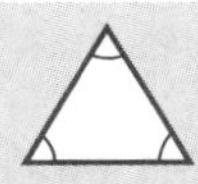

Equivalent equations
Equations that have the same solution.

$3x = 12$, $3x - 12 = 0$, and $4x = 16$ are equivalent equations since the solution for all three equations is $x = 4$.

Equivalent expressions
Expressions that have the same value when simplified.

2(3 + 5) and 2(8) are equivalent expressions since they both have a value of 16.

Equivalent fractions
Fractions that name the same number.

$\frac{2}{4}$ $\frac{3}{6}$ $\frac{4}{8}$

are equivalent fractions as they all represent one half ($\frac{1}{2}$).

Equivalent ratios
Ratios that have the same value.

$\frac{10}{12}$ and $\frac{30}{36}$ are equivalent ratios.

Equivalent sets
Sets containing the same number of elements.

Set A = {1,2,3} and Set B = {a,b,c} are equivalent sets.

Estimate
1. To find an approximate solution to a problem.
2. An answer that is close to the exact answer.

32 x 47 is about 1500.

430 is an estimate for 427.

Euler circuit

A path through a network that includes each edge (arc) exactly once, and that also starts and stops at the same point.

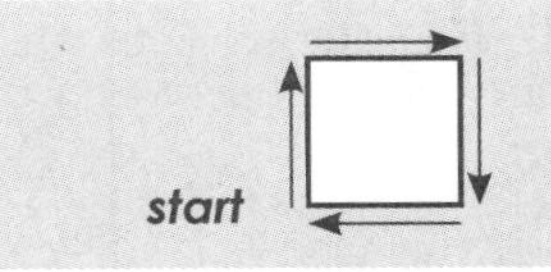

Euler's Formula (Law) for Plane Figures

In a network or plane figure, the relationship among vertices, V, regions, R, and edges, E: **V + R – 2 = E.**

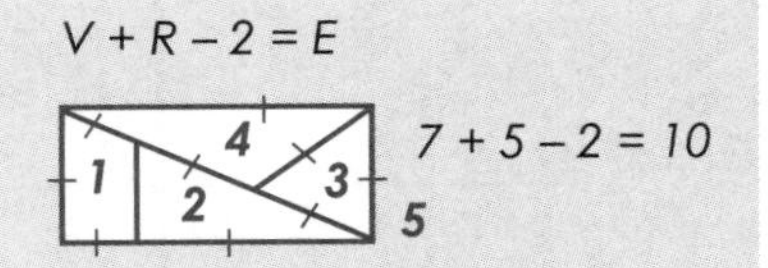

Euler's Formula (Law) for Polyhedra

In a polyhedra, the relationship among vertices, V, regions, R, and edges, E: **V + F – 2 = E.**

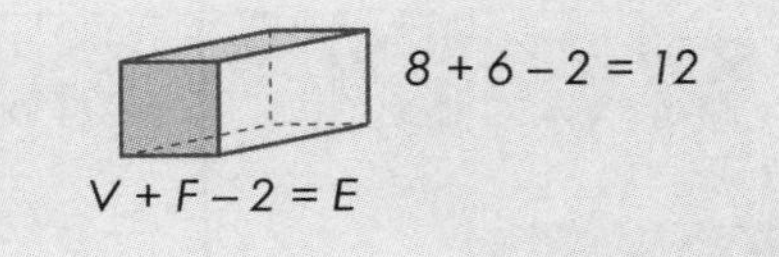

Evaluate

To find the value of an expression with one or more operations.

3(5) + 12 ÷ 3 = 19

Evaluate (an algebraic expression)

To substitute a value for each variable in the expression and simplify the resulting numerical expression.

To evaluate $2x + 4$ when $x = 5$, substitute 5 in the expression, $2(5) + 4 = 14$.

Event

A set of outcomes for an experiment.

The event "getting an even number" when tossing a 6-sided number cube is the set {2,4,6}.

Even number

Any integer that has a reminder of 0 when divided by 2.

0, $^{-}4$, 8, $^{-}32$ and 46 are all even numbers.

Expanded notation

A numerical expression that shows the value or number represented by each digit.

3475 = (3 x 1000) + (4 x 100) + (7 x 10) + (5 x 1)

Expansion

A transformation that increases the size of a figure.

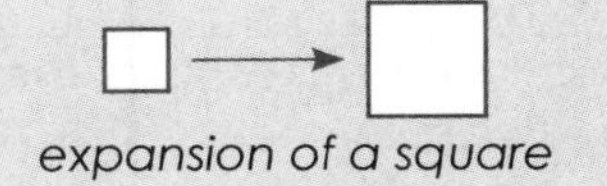

expansion of a square

Experimental probability

A probability based on repeated trials of an experiment. The ratio of the number of times the even occurs to the number of trials.

$$P\ (\text{event}) = \frac{\text{Number of successes}}{\text{Number of trials}}$$

During one month, it rained on 16 of 30 days. The experimental probability that it will rain is $\frac{16}{30}$.

List of Mathematical Terms

Exponent
The number of times a factor is repeated in a power.

The exponent of 3^5 is 5.

Exponential growth (decay)
Growth (decay) that is proportional to size. The larger the quantity gets, the faster it grows.

Bacteria grows exponentially. Radioactive substances decay exponentially.

Expression
See Algebraic expression.

Exterior angle
The supplementary angle to one of the interior angles of a polygon, formed by a side.

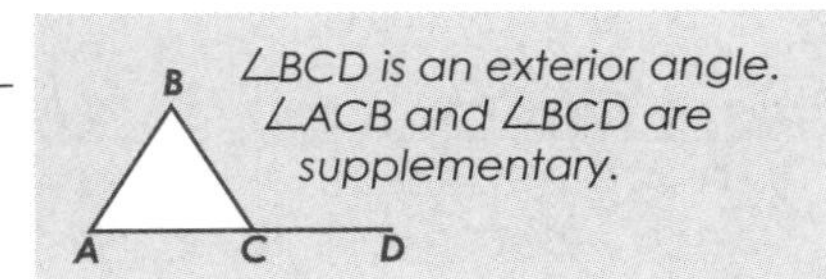

∠BCD is an exterior angle. ∠ACB and ∠BCD are supplementary.

Face
Any 2-dimensional polygonal region that is a side of a polyhedron.

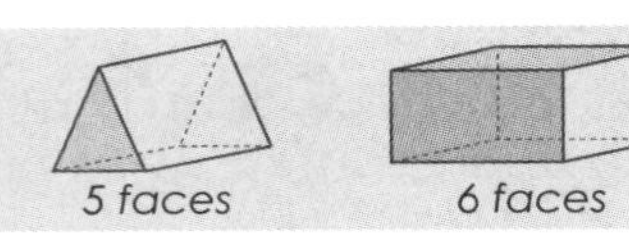

Factor
To write a counting number as a product of counting numbers called factors. The same notion can be applied to integers and algebraic expressions.

Since 2 x 3 x 5 = 30, 2, 3, and 5 are factors of 30.
The integer factors of 4 are ±1, ±2, and ±4.
The factors of $x^2 - x$ are x and $x - 1$.

Factorial (!)
The expression *n!* is read as "*n*-factorial" and represents the product of all integers from 1 to *n*.

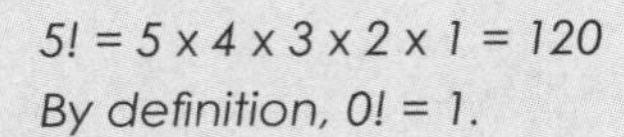

5! = 5 x 4 x 3 x 2 x 1 = 120
By definition, 0! = 1.

Factor tree
A diagram that can be used to write the prime factorization of a number.

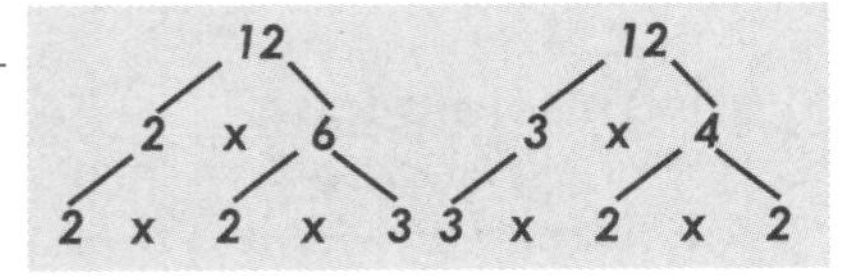

Farenheit (F)
A unit of temperature measurement in the English system with 180 equal divisions between freezing and boiling.

Water freezes at 32 °F and boils at 212 °F.

Favorable outcome
Outcome corresponding to a specified event.

When rolling a pair of 6-sided number cubes, the favorable outcomes for the event "the sum is greater than 10" are (5,6), (6,5) and (6,6).

Fibonacci sequence

The sequence 1, 1, 2, 3, 5, 8, 13, ... where the succeeding numbers in the sequence are generated by adding the previous two. [Discovered by Fibonacci, Leonardo de Piza, in the 13th century.]

Figurate numbers

Numbers that can be represented by arrangements of dots in the shapes of regular polygons.

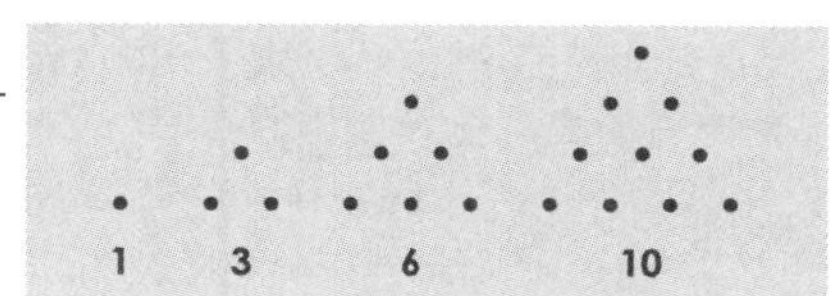

Figure

1. Any 2- or 3-dimensional object or drawing.
2. A number name using digits.

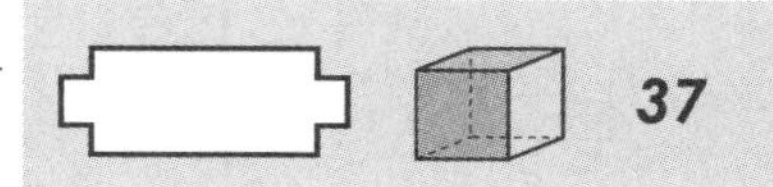

Finite

Has boundaries or can be counted.

First quartile (Q_1)

See Lower quartile.

Flip

A term used for a line reflection transformation.

Foot

Customary (English) system unit of length.

Formula

A mathematical rule, expression, or equation.

$A = l \times w$
$P = 2l + 2w$

Fraction

A number of the form $\frac{a}{b}$, where $b \neq 0$, and a and b are integers.

$\frac{5}{8}, \frac{7}{3}, \frac{9}{1} = 9, 0 = \frac{0}{7} = \frac{0}{-2}, \frac{-6}{7}, \frac{-12}{11}$, and $\frac{-3}{1} = -3$ are examples of fractions.

Frequency

The number of times an event occurs. The number of data values that lie in an interval of a frequency table or histogram.

Class genders

Boys	Girls
14	**16**

Frequency (distribution) table

A table used to display how many times data values or an event occur in a data set.

Interval	Frequency
0 – 9	3
10 – 19	5
20 – 29	6
30 – 39	8
40 – 49	2

Frequency polygon

A polygonal graph formed by joining the midpoints of the tops of the bars of a histogram, which are the midpoints of classes on a frequency table, with an extra class of zero frequency added at each end of the range to complete the graph [see p. 94].

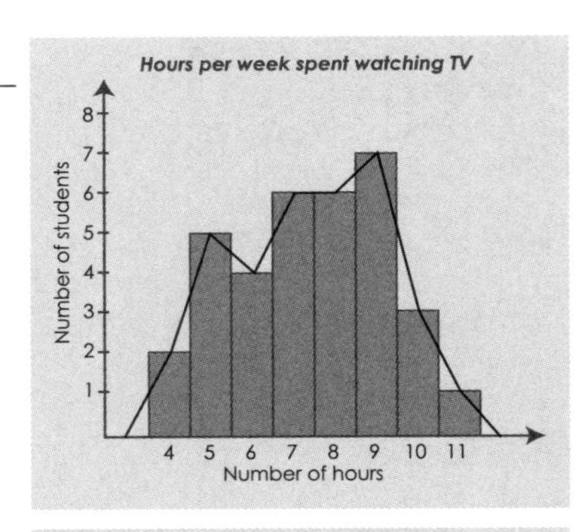

Front-end estimation

A strategy for estimating that first looks at the front-end digits of the numbers, and sometimes the remaining digits are used to adjust the results.

Using front-end estimation for finding the sum 1345 + 4685 + 3247, we obtain 1000 + 4000 + 3000 = 8000, and possibly adjust by adding 1000 to obtain 9000 as a better estimate.

Fundamental Counting Principle

See Counting principle.

Function

A pairing of each number in a given set with exactly one number in another set. A number in the first set is often called an input, and the number in the second set is called the output.

Input	*1*	*2*	*3*	*4*
Output	*3*	*6*	*9*	*12*

Function notation

An expression of the form f(x), where x is the input and f(x) is the output, usually presented as an equation.

$$f(x) = 2x + 5$$
$$f(x) = x^2 - 4x + 1$$

Gallon

A unit of capacity in the English system of measurement.

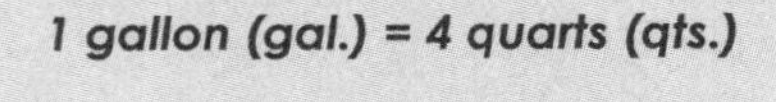

Geoboard

A board with nails or pegs arranged in a pattern and used to represent shapes with rubber bands.

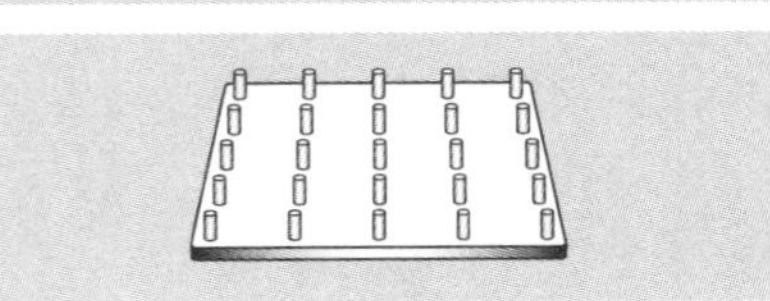

Geometric sequence (progression)

A sequence of numbers where each successive term is obtained from its predecessor by multiplying by a fixed number, called the (common) ratio.

$1, 2, 4, 8, 16, \ldots$

$\frac{1}{3}, \frac{1}{9}, \frac{1}{27}, \frac{1}{18}, \ldots$

$5, 1, \frac{1}{5}, \frac{1}{25}, \frac{1}{125}, \ldots$

Geometry

The study of space and the shapes within it. [From the Greek *geo* for "earth" and *metron* for "measure." The word means "earth measure."]

Glide-reflection

A transformation composed of a translation (glide) and a line reflection.

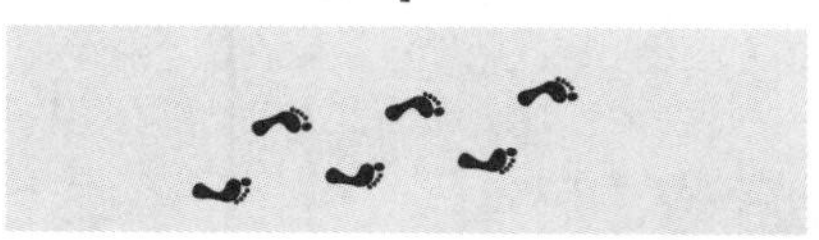

Golden ratio

The classical geometric ratio, used in art and architecture.

$\frac{1+\sqrt{5}}{2} \approx 1.6$

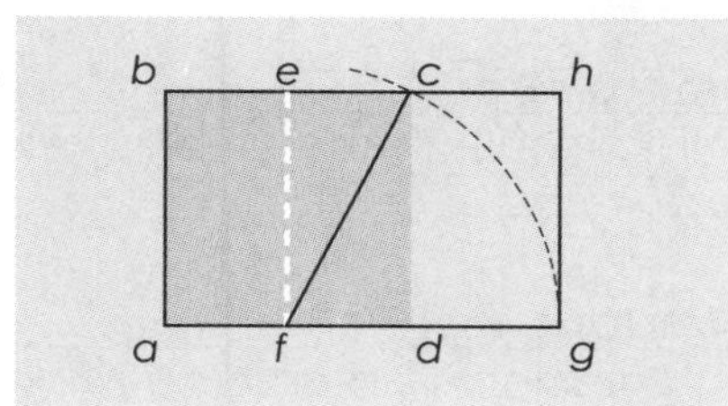

Googol

The number 10^{100} or 1, followed by 100 zeros.

Gram (g)

The basic unit of mass in common usage in the metric system (SI).

The mass of a paperclip is about 1 g.

Graph

On a number line or coordinate plane, the set of points that represent the solution of an equation, inquality, or function.

Great circle

The intersection of a sphere with a plane through the center of the sphere.

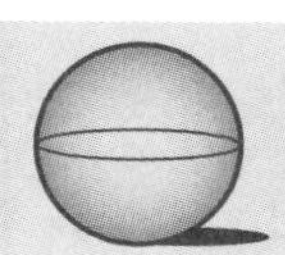

Greater than (>)

Relationship between two numbers or expressions showing which is greater (larger).

$25 > 16, \frac{1}{2} > \frac{1}{4},$

$0.16 > 0.003,$

$3y + 17 > 6y$

Greatest common factor (GCF or gcf)

The greatest (largest) of the common factors of two or more nonzero whole numbers (integers).

The GCF of 18, 27 and 45 is 9 since 9 is the greatest number that is a factor of all three.

List of Mathematical Terms

Grouped data

Data that is organized into groups, according to size and usually of equal interval values; generally used for large amounts of data.

Score	Frequency
1–10	5
11–20	10
21–30	65
31–40	45
41–50	15

Test scores for a large sample are grouped.

Grouping symbols

Symbols such as parentheses, brackets, or fraction bars that group parts of an expression.

$(16 + 8) \div 6$

$\{(2 + 3) \times 5\}$

Half-plane

In a plane, the set of all points on one side of a line.

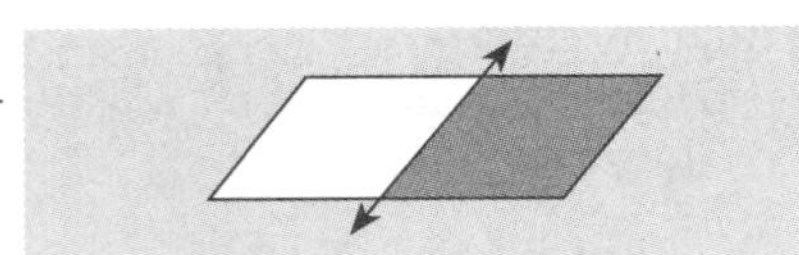

Hectare (ha)

A metric unit of area equivalent to a 100-meter square.

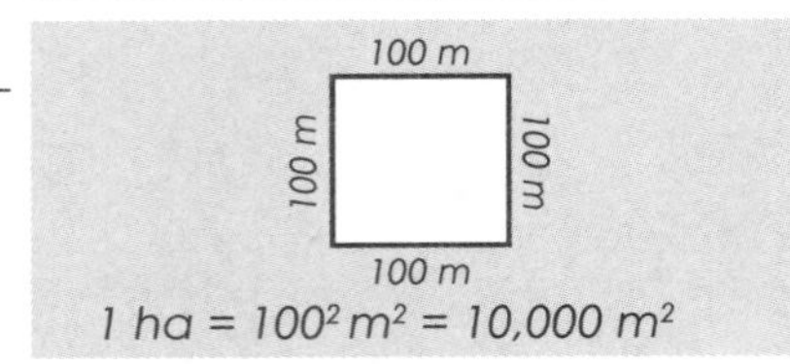

Height (of a parallelogram)

The perpendicular distance between two parallel sides.

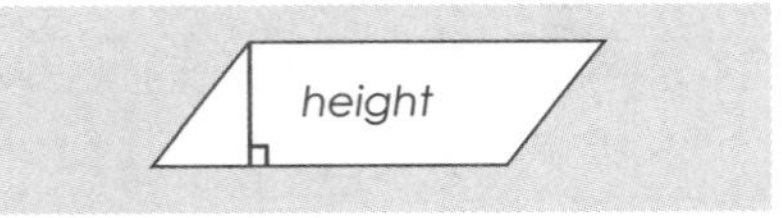

Height (of a trapezoid)

The perpendicular distance between the bases of a trapezoid.

Height (of a triangle)

The perpendicular distance between the base (or the base extended) and the vertex opposite that side.

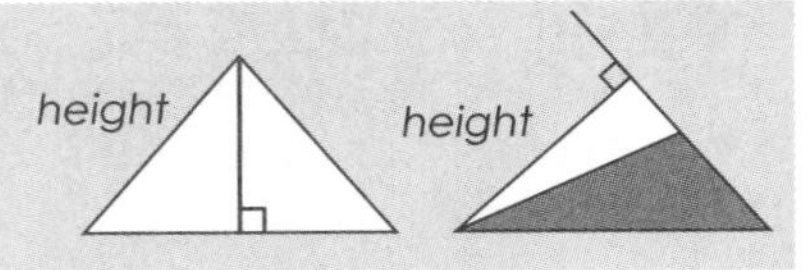

Hemisphere

Half a sphere.

Heptagon (Septagon)

A seven-sided polygon [see p. 99].

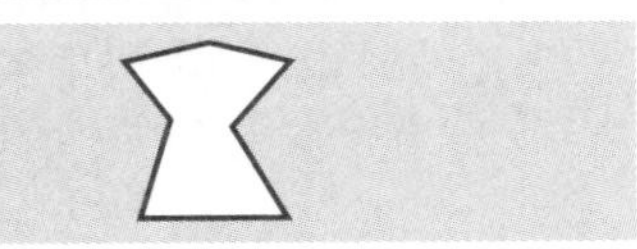

Hexagon

A six-sided polygon [see p. 99].

The cells of a honeycomb are regular hexagons.

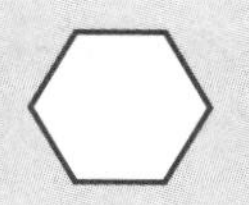

Hexahedron

A polyhedron with 6 faces. A regular hexahedron is one of the Platonic solids.

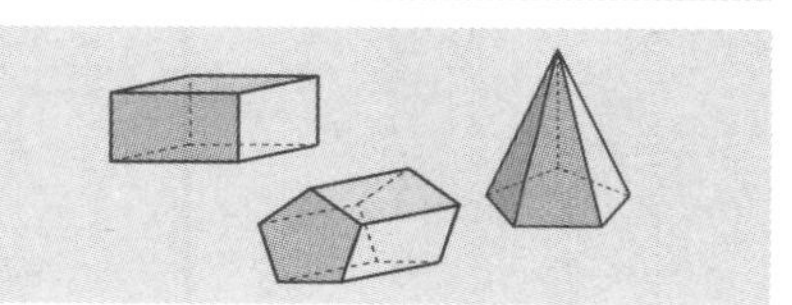

Hindu-Arabic numerals

Our system of numeration is a Hindu-Arabic decimal system (after its region of origin in the 9th century A.D.).

Histogram

A graph that displays data from a frequency table. A histogram has one bar for each interval of the table that contains data values. The length of the bar indicates the frequency of the interval.

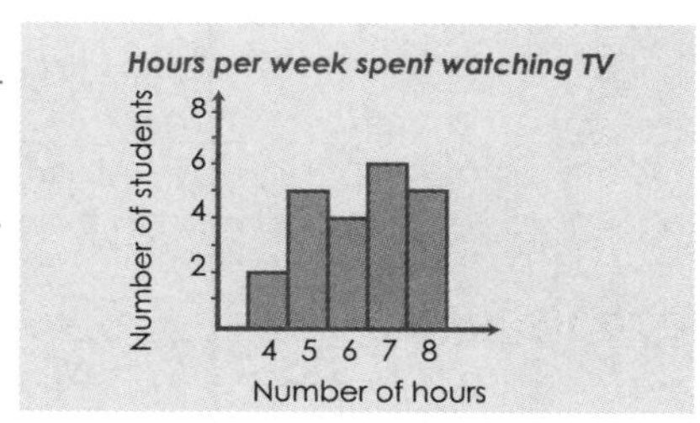

Horizontal

Parallel to the horizon; at right angles to a plumb line (which is vertical).

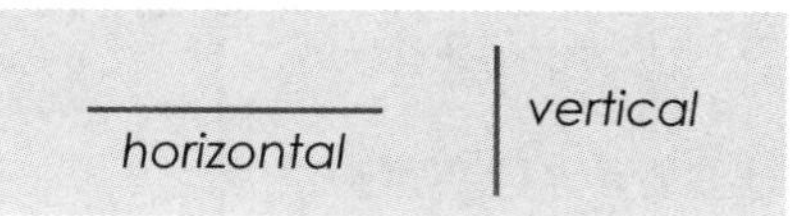

Hour (hr. or h)

An interval of time of 60 minutes.

Hypotenuse

The side of a right triangle that is opposite the right angle.

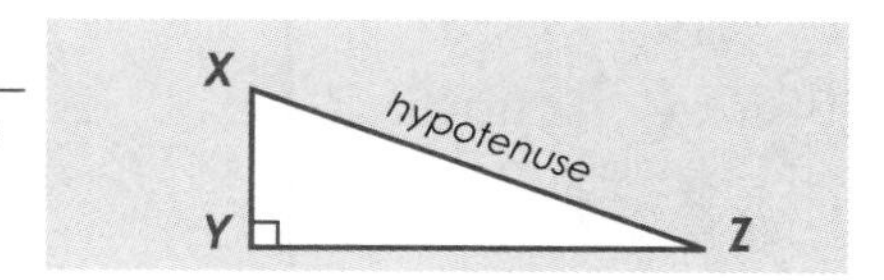

I

Icosahedron

A polyhedron with 20 faces. A regular icosahedron is one of the Platonic solids.

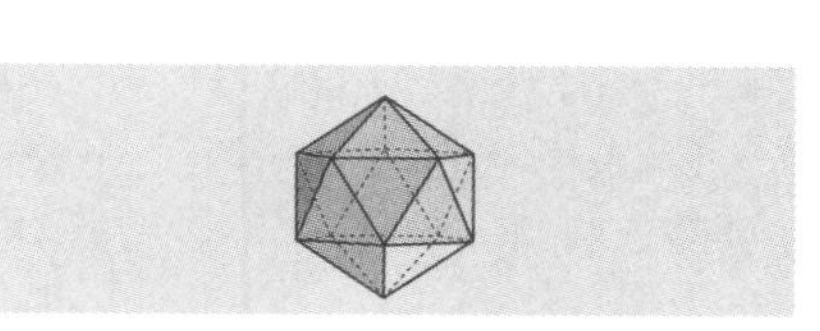

Identity

An equation that is true for all values of the variable(s) included.

$(x + 1)^2 = x^2 + 2x + 1$ is an identity, since it is true for all values of x.

Identity property of additon

See Addition property of zero.

List of Mathematical Terms

Identity property of multiplication

See Multiplication property of one.

Image

The new figure formed as a result of a transformation.

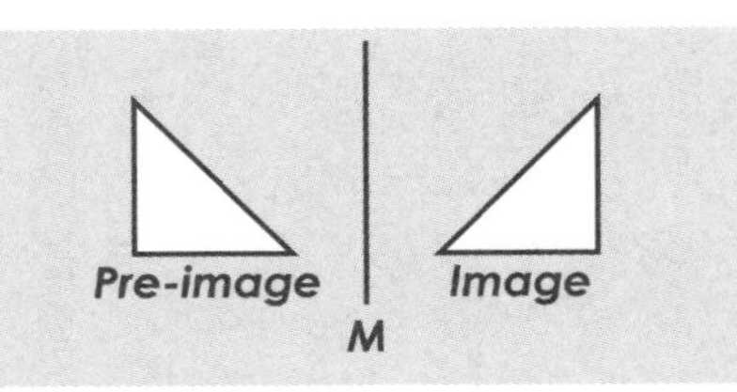

Impossible event

An event that will never happen. An event with probability of zero.

In tossing a six-sided number cube, the event "7" is impossible.

Improper fraction

A fraction, ignoring sign, where the numerator is equal to or greater than its denominator.

$\frac{4}{4}$ $\frac{3}{2}$ $\frac{10}{3}$ $-\frac{10}{3}$ $-\frac{8}{5}$

are improper fractions.

Inch (in. or in)

A customary (English) unit of length, originally established as "the length of 3 barley corns placed end to end."

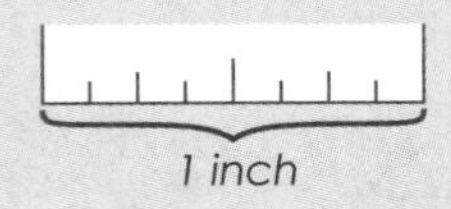

1 inch

Independent events

Two events such that the occurence of one does not affect the likelihood that the other will occur.

When tossing a coin and rolling a number cube, the event "getting heads" and "getting a 5" are independent events.

Index (of a root)

The order of a root. n is the index of $\sqrt[n]{x}$.

In $\sqrt[3]{7}$, 3 is the index.
In $\sqrt{5}$, 2 is understood to be the index

Inequality

A number sentence showing a relationship other than equality.

A mathematical sentence formed by placing an inequality symbol between two expressions.

$4 < 9$, $12 > 1$, $2 + 4 \neq 7$,
$x + 3 < 8$, $2x + 5 \neq 1$

Infinite (Infinity, ∞)

Never ending; not finite.

Counting 1, 2, 3, 4, ... can go on without end; i.e. there is an infinite set of counting numbers.

Input

A number on which a function operates. An input value is in the domain of the function.

Input	*1*	*2*	*3*	*4*
Output	*3*	*6*	*9*	*12*

Inscribed

When a shape is made to fit inside a second shape, the first shape is said to be inscribed in the second.

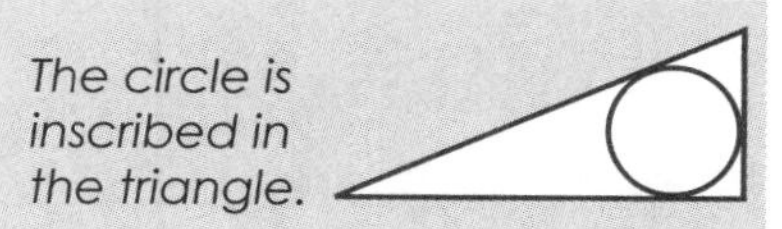

The circle is inscribed in the triangle.

Intercept

The point at which a graph (path) crosses the horizontal x-axis or the vertical y-axis on a coordinate grid.

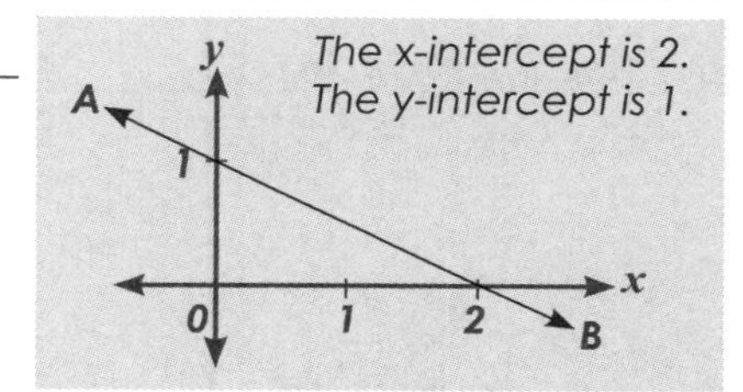

The x-intercept is 2.
The y-intercept is 1.

Integers (I)

The set of counting numbers, their opposites, and 0. I = {…, –2, –1, 0, 1, 2, …}

4 and –13 are integers. $-4\frac{1}{2}$ *and 23.8 are not integers.*

Interest

The amount earned or paid for the use of money.

If $100 is invested annually at a rate of 5%, then the amount of interest is $5.

Interest rate

The percent used to determine the interest.

See above example.

Interior angles (of a polygon or polyhedron)

The angles interior to the sides of the polygon or polyhedron.

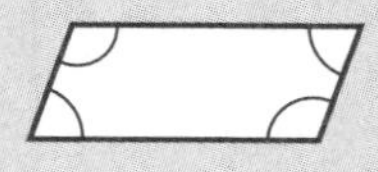

Intersecting lines

Lines that share exactly one point.

Intersection

1. The point at which two or more paths cross.
2. The elements common to two or more sets.

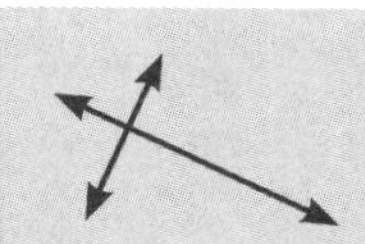

If A = {1, 2, 3, 4, 5} and B = {2, 4, 6, 8}, then the intersection of A and B is $A \cap B = \{2, 4\}$.

List of Mathematical Terms

Interval

The amount of distance or time between two positions or events; or the range between two given numbers.

- *An interval of 5 mm between two points on a ruler.*
- *An interval of 5 minutes between the start and finish of a race.*

Inverse

See Additive inverse, Multiplicative inverse.

Inverse operations

Operations that "undo" each other.

Addition and subtraction are inverse operations:
$3 + 5 = 8$ and $8 - 5 = 3$.
Multiplication and division are inverse operations:
$4 \times 5 = 20$ and $20 \div 5 = 4$.

Inverse proportion

Two quantities are in inverse proportion when one quantity increases at the same rate as the other decreases, and vice versa. Note that the product of two such quantities is constant.

Speed (km/h)	Time (hours)
60	1
30	2
20	3
15	4
12	5
10	6

Irrational numbers (Q')

Real numbers that cannot be written as a quotient of two integers. A real number that cannot be expressed as a terminating or repeating decimal.

$\sqrt{2}, \pi, 3 - \sqrt{7}$

0.12112111211112 ...

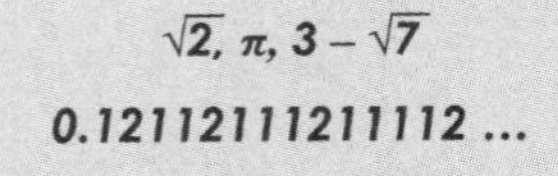

Isometric paper

Paper with a pattern of line segments or dots with constant lengths or intervals. [From the Greek *isos* for "equal" and *metron* for "measure."]

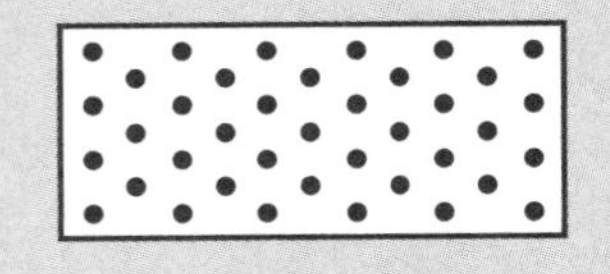

Isosceles triangle

A triangle with at least two sides of the same length. An equilateral triangle is also isosceles. [From the Greek *isos* for "equal" and *skelos* for "leg," thus "equal-legged"; see p. 100.]

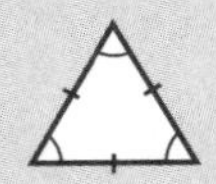

Isosceles trapezoid

A trapezoid with exactly one pair of non-adjacent congruent sides.

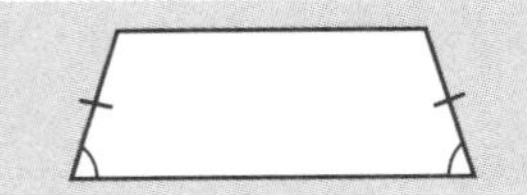

J

Joule (J)

A unit of energy or work, such that 1 J is the amount required to raise 1 g (1 mL) of water 1 °C. [Replaces the calorie unit and named after its developer, James Joule.]

K

Kilo (k)

Prefix for one thousand in the metric system (SI).

Commonly used units are kilogram (kg), kiloliter (kL), kilojoule (kJ), and kilometer (km).

Kite

A quadrilateral with two distinct pairs of adjacent sides congruent.

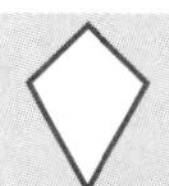  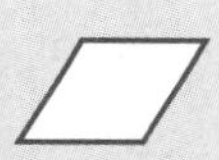

All of these quadrilaterals are kites.

Knot (kn)

A unit of speed of one nautical mile per hour, usually applied to wind or boat speeds. [A nautical mile is approximately 1.85 km; see p. 123.]

L

Lateral face

In a prism or a pyramid, a face that is not a base.

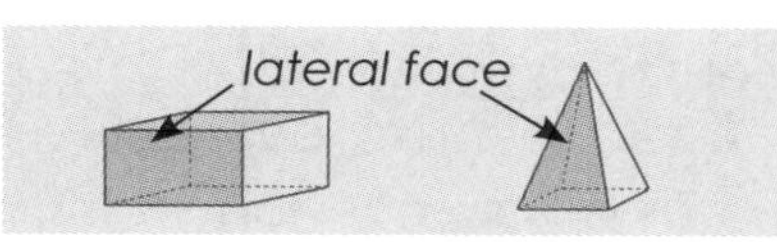

Latitude

An imaginary line joining points on the Earth's surface that is of equal distance north or south of the equator. See Longitude.

Leading digit

The first nonzero digit in a number.

The leading digit of 642 is 6.
The leading digit of 0.003875 is 3.

Leaf

The last digit on the right of a number displayed in a stem-and-leaf plot.

Stem	Leaves
1	9 6 9
2	1 7 6 4 8
3	4 5 2

A stem-and-leaf plot for data 21, 19, 34, 16, 27, 35, 19, 26, 24, 28, 32. Examples of leaves are 9, 1 and 4.

Leap year

A year with 366 days and normally occurring every four years; i.e. 2008, 2012, 2016, ... are leap years (the extra day being February 29). [Note that century years must be divisible by 400 to be leap years, so that 2000 was a leap year but 2100 will not be a leap year; see p. 122.]

List of Mathematical Terms

Least (lowest) common denominator (LCD or lcd)

The least common multiple of the denominators of two or more fractions.

The least common denominator of $\frac{5}{6}$ and $\frac{4}{9}$ is the least common multiple of 6 and 9, which is 18.

Least (lowest) common mutiple (LCM or lcm)

The smallest of the common multiples of two or more nonzero whole numbers.

The least common multiple of 3 and 4 is the smallest of the common mulitples 12, 24, 36, 48, 60, ..., which is 12.

Legs (of a right triangle)

The two sides of a right triangle that form the right angle.

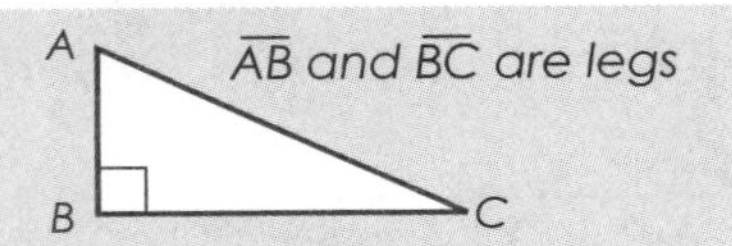

Length

The measure of a path or object in one dimension from end to end. See Dimension.

The length of this line segment is 4 cm.

Less than (<)

A relation between two numbers or expressions showing which is smaller.

$27 < 30$, $\frac{1}{3} < \frac{1}{4}$, $0.13 < 0.24$, $3y < 3y + 1$

Like terms

Terms that have identical variable parts.

Two or more constant terms are also considered like terms.

In the algebraic expression, $3x^2 - 5 + y - y^2 - x^2 + 3y + 6$, the terms $3x^2$ and x^2, y and $3y$, and -5 and 6 are like terms.

Line

A set of points that extends without end in two opposite directions.

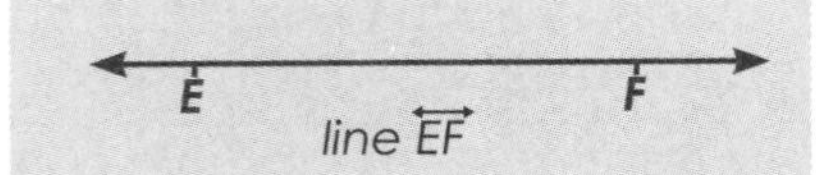

Linear equation

An equation whose graph is a line. [Note that in linear equations there are only one or two variables, usually represented by x and y, and the variables are raised to only the first power.]

$y = 3$

$x + y = 5$

are linear equations.

Linear function

A function whose graph is a line or part of a line.

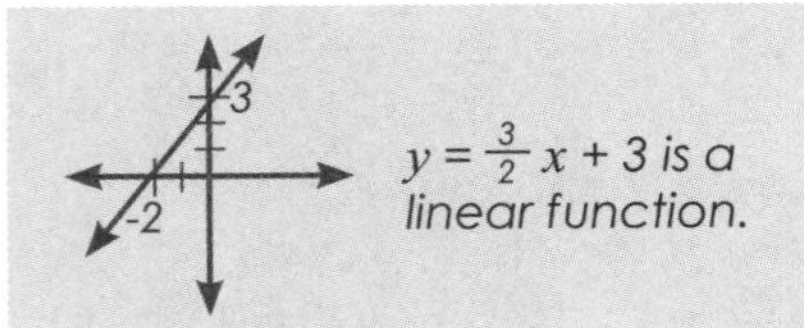

$y = \frac{3}{2}x + 3$ *is a linear function.*

Linear inequality

An inequality in which each term contains at most one variable to the first power.

$3y < 2x + 1$ is a linear inequality.

Line graph

A type of graph in which points representing data pairs are connected by line segments [see p. 94].

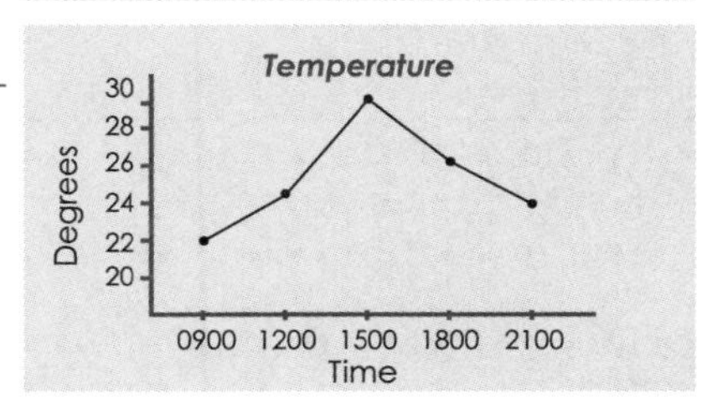

Line of reflection

The line over which a figure is reflected (flipped) when the figure undergoes a reflection transformation.

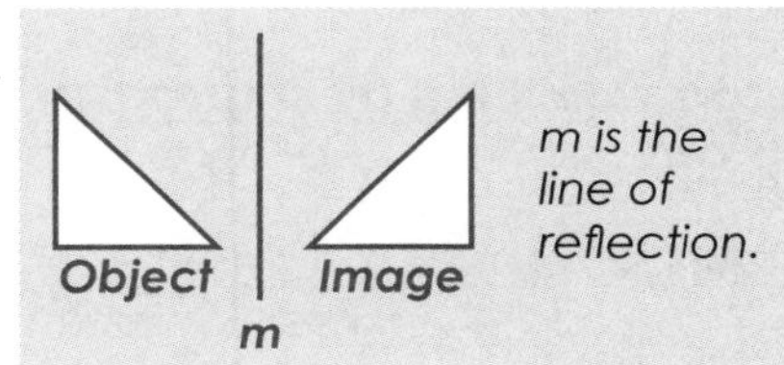

Line segment

A subset of a line that contains two points of the line and all the points between these two points.

An open line segment does not contain its endpoints.

A half-open (half-closed) line segment contains only one of the endpoints.

line segment: A B $\overline{AB}$

open line segment: A B $\overset{\circ\!-\!\circ}{AB}$

half-open line segment: A B $\overset{\circ\!-\!\bullet}{AB}$ A B $\overset{\bullet\!-\!\circ}{AB}$

Line of symmetry (Axis of symmetry)

A line that divides a figure into two parts that are mirror images of each other.

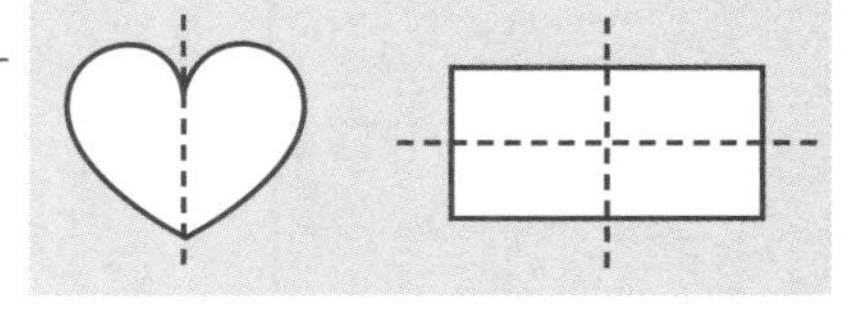

Line plot

A number line diagram with marks or dots to show frequencies.

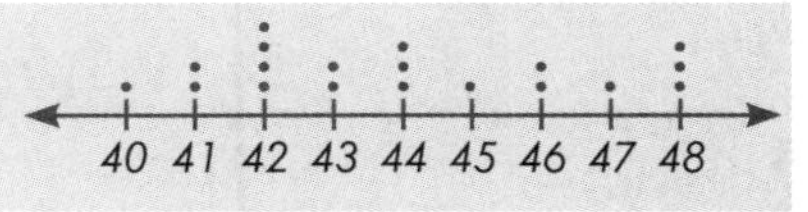

Line symmetry (Reflective symmetry)

A figure has line symmetry if it can be divided by a line, called a line of symmetry, into two parts that are mirror images of each other.

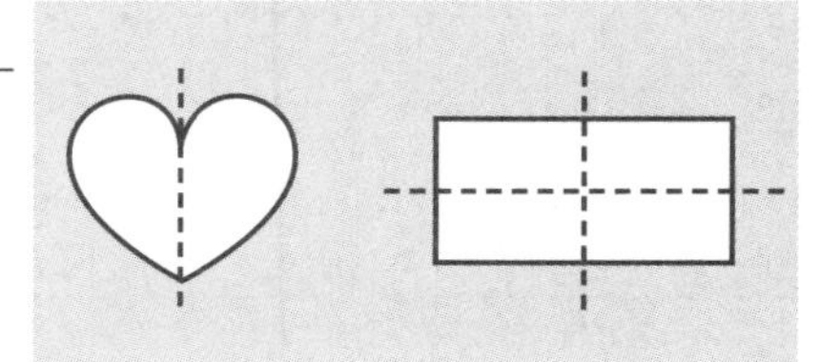

List of Mathematical Terms

Liter (L)

The basic unit of capacity for liquids and gases in the metric system.

A liter of milk.

Longitude

The angular distance east or west of the imaginary line (prime meridian) that stretches from the North Pole to the South Pole and passes through Greenwich, England. See Latitude.

Lower extreme (value)

The least value in a data set.

Lower quartile (First quartile, Q_1)

The median of the lower half of a data set. As a percentile, P_{25}.

Lowest terms

An outdated expression referring to simplest form (simplest terms) of a fraction. See Simplest terms.

M

Magic square

An arrangement of numbers in a square grid so that the sums of every row, column and diagonal are the same.

4	9	2
3	5	7
8	1	6

Magnitude

The size of a quantity or number.

Major arc

The longer of two arcs connecting two points on a circle.

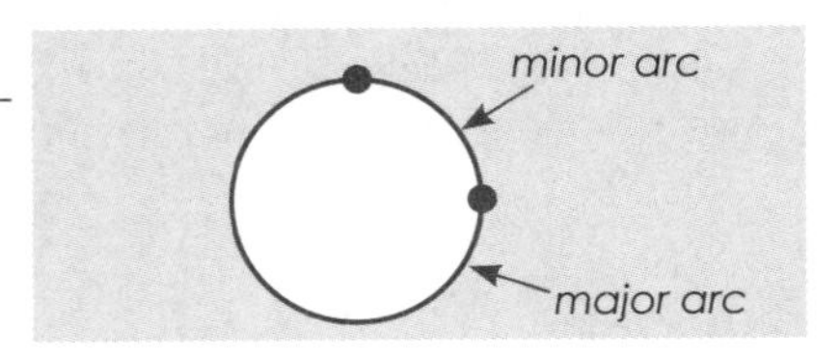

Markup

The increase in the wholesale price of an item.

The wholesale price of an item is $1, and a store sells it for $1.25. The markup is $0.25.

Mass

Amount of matter, commonly measured in grams, kilograms and metric tons.

List of Mathematical Terms

Maximum

The highest number, or greatest size or amount.

The maximum speed legally allowed on many roads is 55 mph.

Mean ($\bar{x}$) (Arithmetic mean)

The sum of the values in a data set divided by the number of values.

The mean of the data set 72, 75, 81, 92 is **(72 + 75 + 81 + 92) ÷ 4 = 80.**

Measure

1. A number assigned to a quantity to indicate its size compared to a chosen unit.

   Choosing a centimeter unit, the width of this page is about 17 cm.

2. To find dimensions, capacity, weight, or other quantities.

   We usually measure angles with a protractor in degree units. 30°

Measures of central tendency

Measures used to describe the middle of a data set. See Mean, Median and Mode.

Median

The middle value or the average of the two middle values in an ordered data set.

In the set 8, 10, 12, 14, 16, the median is 12.
In the data set 8, 10, 12, 14, the median is (10 + 12) ÷ 2 = 11.

Mega

Metric prefix for one million.

Megabytes (Mb) are used to measure computer storage space.

Meter (m)

The basic unit of length in the metric system. Originally defined as one forty-millionth of the circumference of the Earth.

Metric system

A decimal system of measurement used internationally, and from which the SI (System Internationale) is derived [see p. 118].

Micro (m)

Metric prefix for $\frac{1}{1,000,000}$. Note: the abbreviation for micro and milli are both m.

Time is often measured in microseconds (ms).

Midpoint

The point which divides a line segment into two equal parts.

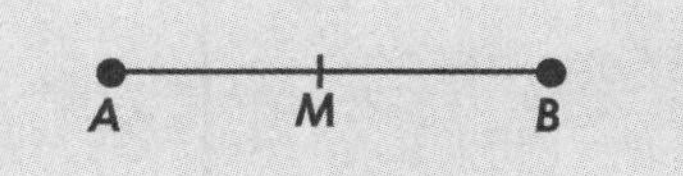

List of Mathematical Terms

Mile

A unit of length in the customary (English) system.

1 mile = 5280 feet
1 mile ≈ 1.6 km

Millennium

A period of a thousand years.

We are now in the third millennium A.D.

Milli (m)

The metric prefix for $\frac{1}{1000}$.

Commonly used units in the metric system are millimeter (mm), milligram (mg), milliliter (mL), and millisecond (ms).

Million

The number 1,000,000 = 10^6.

Minimum

The smallest number, size, or amount.

The minimum temperature in the city last night was 1 °F.

Minor arc

The shorter of two arcs connecting two points on a circle.

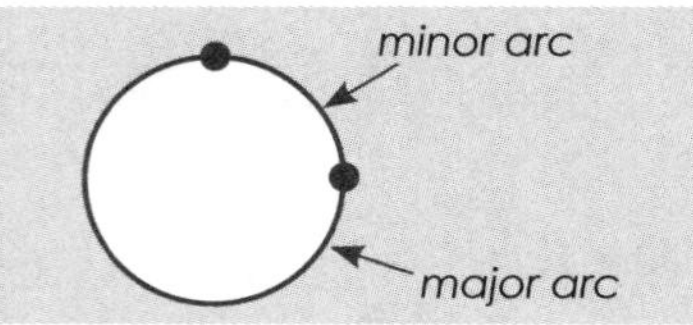

Minus (–)

1. Subtract or take away.
2. The symbol used to show a negative amount.

Ten minus three is seven; or 10 – 3 = 7.

⁻2, read as "negative 2."

Minute

One-sixtieth of both a unit of time of one hour, and a unit of angle measure of one degree. [Originates from the ancient Babylonians who used a base of 60 to represent numbers; see p. 121 and p. 123.]

Mixed number (numeral)

A number that has a whole number part and a fraction part.

$15\frac{3}{4}$ and $37\frac{1}{2}$ are mixed numerals.

Mode

The value in a data set that occurs most often. A data set can have no mode, one mode, or more than one mode.

In the set of scores 6, 6, 7, 8, 8, 8, 9, 10, 10, the mode is 8.

Monomial

A number, a variable, or a product of a number and one or more variables.

$2x$, 6, $5xy$, x^2, and $8x^2y^3$ are monomials.

Month

A period of time—a year is divided into 12 months [see pp. 121–122].

Multi-base Arithmetic Blocks (MAB)

Wooden, plastic or foam blocks to represent numbers in a variety of bases. See Base ten blocks.

Multiple

The product of a given counting number and any whole number. This concept may be extended to include integer multiples.

The multiples of 3: 0, 3, 6, 9, … or the (integer) multiples of 3: …, -9, -6, -3, 0, 3, 6, 9, …

Multiplicand

In the product $a \times b$, a is called the multiplier (factor); b is called the multiplicand (factor).

multiplicand

$4 \times 9 = 36$ or $\begin{array}{r} 9 \\ \times\, 4 \\ \hline 36 \end{array}$

multiplier product

Multiplication (x)

Repeated addition, calculated by operating on two numbers to find their product. See also Cartesian product and Array.

$3 + 3 + 3 + 3 = 4 \times 3 = 12$

Multiplication property of equality

Multiplying each side of an equation by the same nonzero number produces an equivalent equation.

If $\frac{x}{4} = 6$, then $(4)(\frac{x}{4}) = (4)(6) = 24$, or $x = 24$.

Multiplication property of zero (Zero property of multiplication)

The product of any number and zero is zero [see p. 72].

$16 \times 0 = 0 \times 16 = 0$
$a \times 0 = 0 \times a = 0$

Multiplicative identity

The number 1 is the multiplicative identity since the product of any number and 1 is the original number.

$16 \times 1 = 1 \times 16 = 16$
$a \times 1 = 1 \times a = a$

Multiplicative inverse

The multiplicative inverse of a nonzero number $\frac{a}{b}$, is the reciprocal of the number, or $\frac{b}{a}$. The product of a number and its multiplicative inverse is 1.

The multiplicative inverse of $\frac{3}{5}$ is $\frac{5}{3}$. The multiplicative inverse of 7 is $\frac{1}{7}$.

$\frac{3}{5} \times \frac{5}{3} = 1$; $7 \times \frac{1}{7} = 1$

List of Mathematical Terms

Multiplier

In the product $a \times b$, a is called the multiplier (factor); b is called the multiplicand (factor).

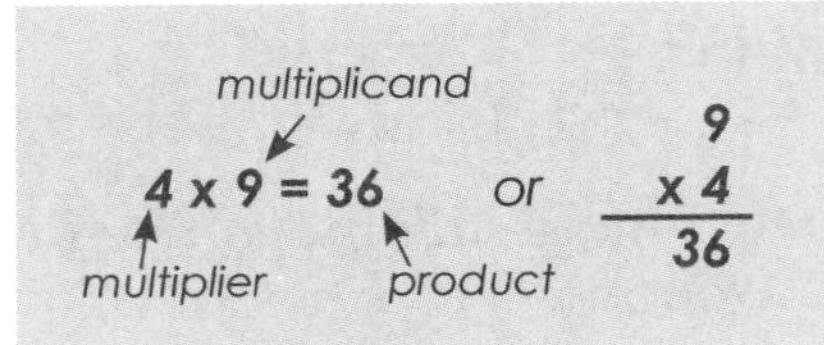

Mutually exclusive events

Events that have no outcomes in common.

When rolling a 6-sided number cube, the events "roll an even" and "roll an odd" are mutually exclusive.

N

Nano (n)

Metric prefix for $\frac{1}{1,000,000,000}$ (one billionth).

Nanoseconds (ns) are used to measure time.

Natural numbers (N)

The natural numbers are the same as the counting numbers.

N = {1, 2, 3, 4, 5, ...}

Nautical mile

A unit of length around the Earth's surface determined by the arc formed by an angle of measure 1 minute ($^1/_{60°}$) at the Earth's center (used in navigation).

1 nautical mile ≈ 1.85 km

Negative integers (I^-)

The integers less than zero. $I^- = \{..., -3, -2, -1\}$

-2 and -5120 are negative integers.

Negative numbers

The real numbers less than zero.

-4, $-\frac{2}{3}$, -1.6, $-\pi$, $-\sqrt{7}$, and $-.007$ are negative numbers

Net

A 2-dimensional figure that can be folded to form a 3-dimensional shape.

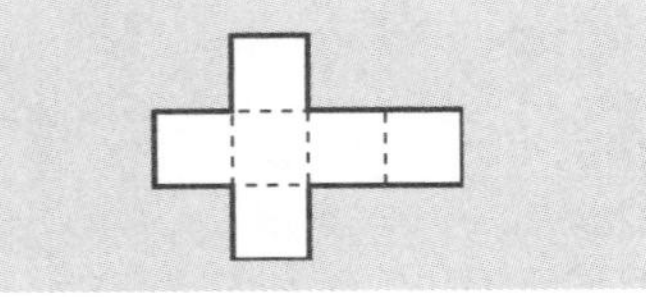

Network

A system consisting of points (vertices/nodes) and curves or line segments (edges, arcs) that connect pairs of points.

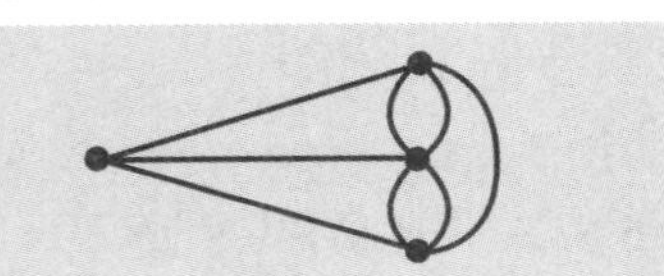

Node (Junction)

A point (vertex) in a network.

List of Mathematical Terms

Nominal number
A number that names a person or thing.

Player #33, Chapter 10

Nonagon
A nine-sided polygon [see p. 99].

Not equal (≠)
Two numbers or quantities that are not the same are said to be not equal.

$6 + 5 \neq 10$

$2 \text{ cm} \neq 21 \text{ mm}$

Number fact family
Four number facts consisting of 3 numbers related by inverse operations.

Under addition/subtraction:
$5 + 7 = 12$, $7 + 5 = 12$
$12 - 5 = 7$, $12 - 7 = 5$

Under multiplication/division:
$2 \times 3 = 6$, $3 \times 2 = 6$
$6 \div 3 = 2$, $6 \div 2 = 3$

Number line
A line whose points are associated with numbers. The numbers on a number line increase from left to right.

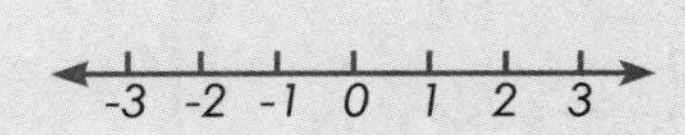

Number pattern
An organized arrangement or sequence of numbers.

0, 5, 10, 15, 20, ...

Number sentence
An equality or inequality concerning numbers. A number sentence can be true or false.

$12 \div 5 = 2.4$ and $16 > 5 \times 3$ are both true; the sentences $10 \div 3 = 4$ and $4 > 7$ are both false.

Number theory
A branch of mathematics dealing with the natural numbers.

Numeral (Number name)
One or more symbols used to represent a number. [From the Latin *enumerate* meaning "to count out"; see pp. 73–74.]

XV and 15 are two numerals which represent the number fifteen.

Numeration system
A collection of symbols and rules used to represent numbers.

Our numeration system is the Hindu-Arabic system with ten digits and base ten place value.

List of Mathematical Terms

Numerator

The number a in the fraction $\frac{a}{b}$.
The number above the fraction line (vinculum) which tells how many of the named fraction are being considered.

In the fraction $\frac{3}{4}$ the numerator is 3, indicating that there are 3 fourths.

Numerical expression

An expression that contains only numbers and operations and that represents a particular value.

The numerical expression 5 + (3 x 4) – 1 represents 16.

Oblique

Slanting; any shape that is neither vertical nor horizontal is in an oblique position.

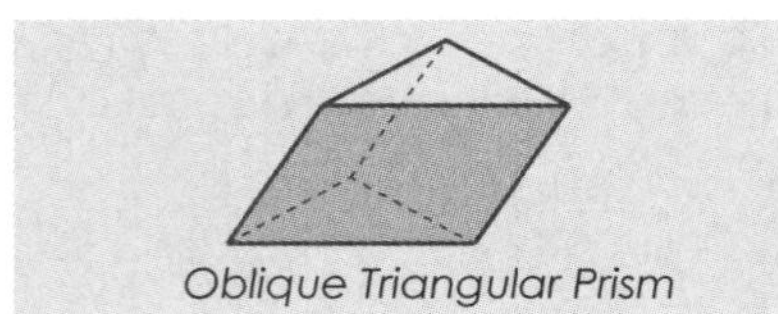

Oblique Triangular Prism

Oblong

Not square.

Oblong number

The product of two consecutive counting numbers.

2, 6, 12, and 20 are oblong numbers.

Obtuse angle

An angle whose measure is between 90° and 180°.

Angle EFG is an obtuse angle.

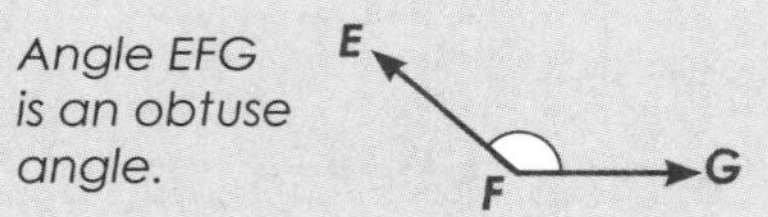

Obtuse triangle

A triangle with one obtuse angle.

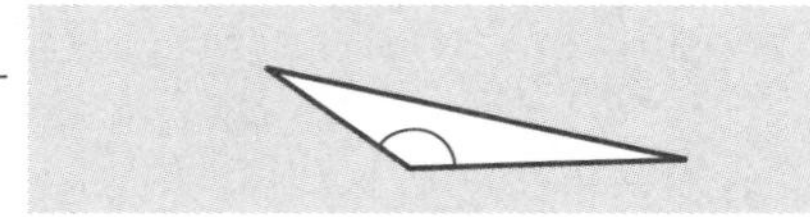

Octagon

An eight-sided polygon [see p. 99].

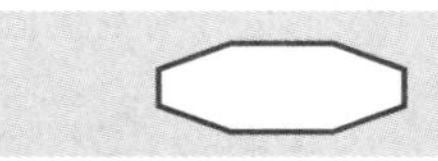

Octahedron

A polyhedron with eight faces [see p. 109].

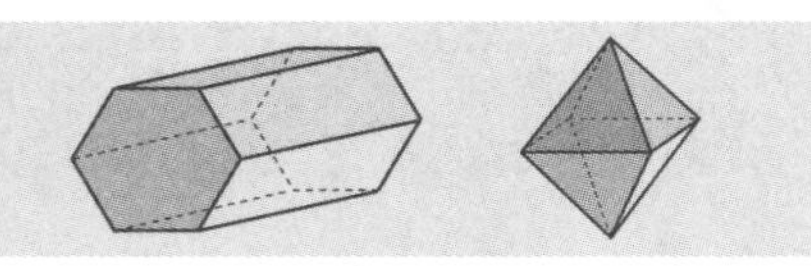

Odd numbers (O)

Whole numbers not evenly divisible by 2.
O = {1, 3, 5, 7, 9, …}
The odd integers include the opposites of the odd numbers.

39 and 379 are odd numbers.
-1, 39 and -379 are odd integers.

Odds (in favor of an event)
The ratio of the probability that an event occurs to the probability that it does not occur.

The odds of rolling a number less than 3 on a six-sided number cube are $\frac{2}{4} = \frac{1}{2}$, or 1:2.
The odds of getting 2 heads when tossing two coins are $\frac{1}{3}$ or 1:3.

One-dimensional (Linear) (1-D)
A path or line having only length.

Both figures shown here are 1-D.

One-to-one correspondence (1-1 correspondence)
A pairing of the members of one set with the members of another set so that each member has exactly one partner.

If $A = \{a, b, c\}$ and $B = \{1, 2, 3\}$, then $a \rightarrow 1$, $b \rightarrow 2$, and $c \rightarrow 3$ is a 1–1 correspondence.

Open number sentence
A mathematical sentence containing one or more variables.

$6 + y = 10$; $z \times z = 25$; $10 - w = 9$ are open sentences.

Operation (binary operation)
A defined process for combining two numbers to create another number.

Addition	**$3 + 5 = 8$**
Subtraction	**$8 - 5 = 3$**
Multiplication	**$2 \times 4 = 8$**
Division	**$8 \div 4 = 2$**
Exponent	**$2^3 = 8$**
Root	**$\sqrt[3]{8} = 2$**

Opposites (additive inverses)
Two numbers that are the same distance from 0 on a number line but are on opposite sides of 0. Zero is its own opposite.

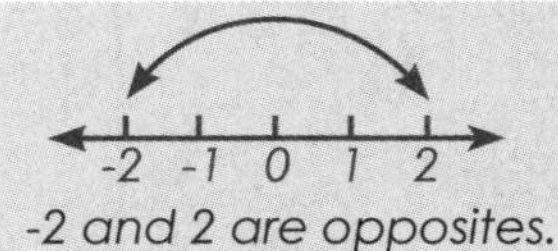

-2 and 2 are opposites.

Order
To place in a list according to an attribute such as size, color, or numerical value.

21 18 15 12 9
The numbers shown are in descending order from left to right.

Ordered pair
A pair of numbers (x, y) that can be used to represent a point in a coordinate plane. The first number is the x-coordinate, also referred to as the abscissa. The second number is the y-coordinate, also referred to as the ordinate.

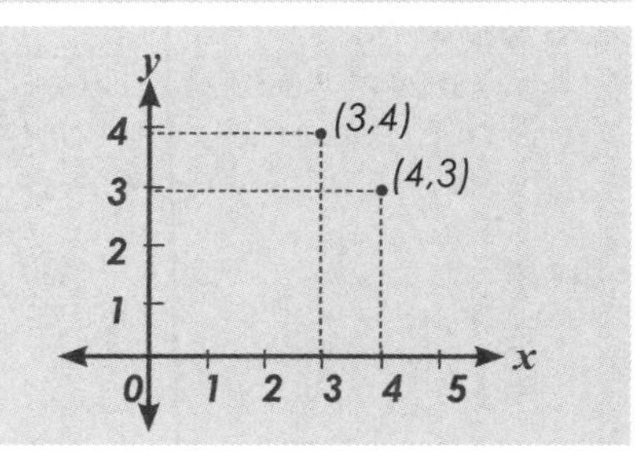

List of Mathematical Terms

Order of operations

The order in which to perform operations when evaluating expressions with more than one operation.

Operations are performed in the order given by the pneumonic PEMDAS:

1. P—Parentheses (and brackets)
2. E—Exponents (and roots)
3. MD—Multiplication and/or Division (left to right)
4. AS—Addition and/or Subtraction (left to right)

See the steps in evaluating the following expression:

$20 \div 5 \times 6 + (16 - 3) - 3 \times 2^3 + 9$ **[P]**

$= 20 \div 5 \times 6 + 13 - 3 \times 2^3 + 9$ **[E]**

$= 20 \div 5 \times 6 + 13 - 3 \times 8 + 9$ **[MD]**

$= 24 + 13 - 24 + 9$ **[AS]**

$= 22$

Ordinal number

A number used to indicate the position of an object in an ordered sequence.

1st, 2nd, 3rd, 4th etc.

Ordinate

See *y*-coordinate.

Origin

The point (0, 0) where the x-axis and the *y*-axis meet in the coordinate plane.

The point (0, 0, 0) where the x-, *y*- and z- axes meet in 3-dimensional space.

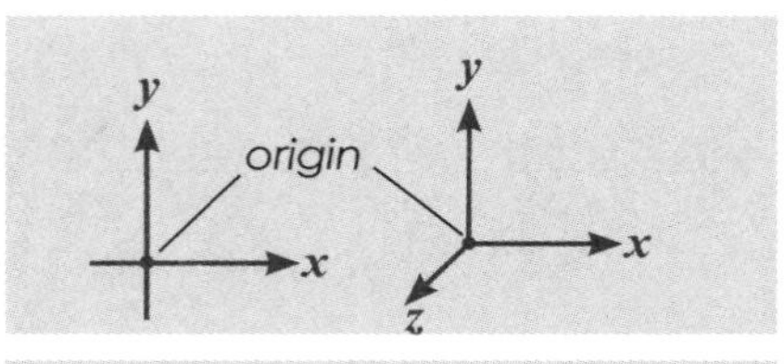

Ounce (oz.)

A unit for measuring weight in the English system.

16 ounces (oz.) = 1 pound (lb.)

Outcome

A possible result of an experiment.

When tossing a coin, the outcomes are heads and tails.

Outlier

A data point that is significantly greater than or less than the main body of data points.

In the set of scores 2, 11, 11, 12, 12, 13, 14, 14, 14, 15, the score of 2 is an outlier.

Output

A number produced by evaluating a function using a given input. An output value is in the range of the function.

Input	***1***	***2***	***3***	***4***
Output	***3***	***6***	***9***	***12***

Oval

See Ellipse.

Overlapping events

Events that have one or more outcomes in common; events that are not mutually exclusive.

When rolling a number cube, the events "getting a number greater than 4" and "getting an even number" are overlapping events because they have the outcome 6 in common.

Ovoid

Egg-shaped object or figure. [From ovum meaning egg or ova (plural).]

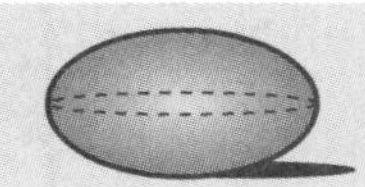

P

Palindrome

A number name, date, word, or sentence that reads the same both forwards and backwards.

747	***01–02–2010***
level	***Madam I'm Adam***

Parabola

A curve obtained by cutting a cone with a plane; one of the conic sections.

The cross-section of a headlight reflector, as shown.

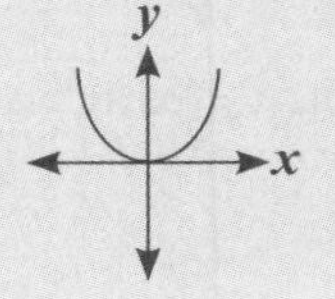

In the coordinate plane, $y = x^2$ describes a parabola with vertex (0, 0).

Parallel lines (II)

Two lines in the same plane that do not intersect. Two lines in the same plane that are everywhere equidistant.

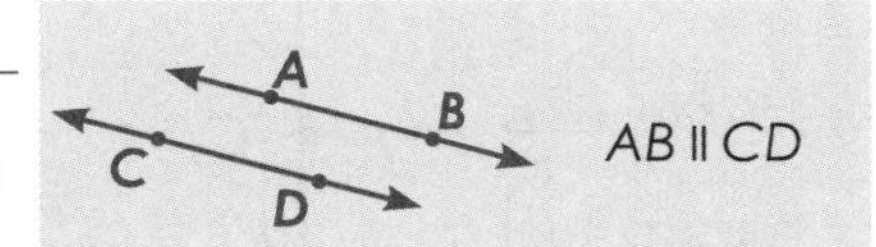

Parallel planes

Two planes that do not intersect. Two planes that are everywhere equidistant.

The top and bottom of the box lie in parallel planes.

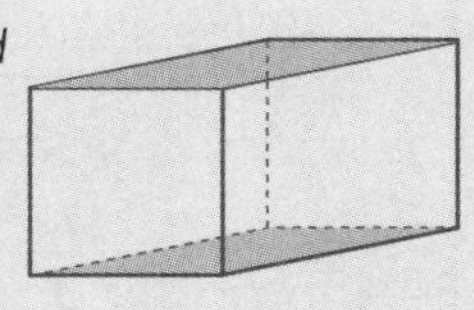

Parallelogram

A quadrilateral with two pairs of parallel sides which are also congruent.

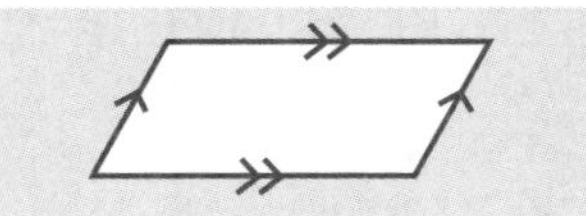

Partition

To divide or "break up."

128 could be partitioned as 100 + 20 + 8, or as 32 + 32 + 32 + 32, etc.

List of Mathematical Terms

Pascal's Triangle

A triangular arrangement of numbers in which each row starts and ends with 1, and each number is the sum of the two numbers above it. These numbers are the binomial coefficients, used in counting and probability. [Blaise Pascal was a famous French mathematician who lived in the 17th century.]

1 1
1 2 1
1 3 3 1
1 4 6 4 1
1 5 10 10 5 1
1 6 15 20 15 6 1

Note: 1 was not placed at the top of the triangle as it was not part of Pascal's Triangle.

Pattern

A sequential list of objects or numbers with a consistent relationship between them.

1, 2, 3, 1, 2, 3, 1, 2, 3, 1, 2, 3

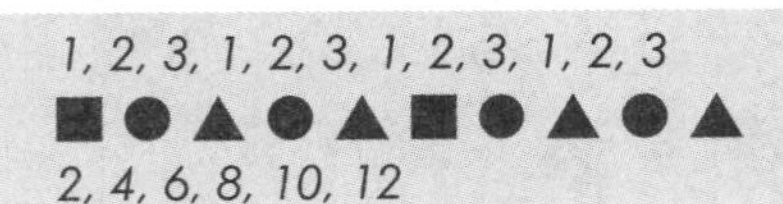

2, 4, 6, 8, 10, 12

Pattern blocks

Sets of plastic, wood, or foam shapes in the form of triangles, rhombuses, squares, trapezoids and hexagons.

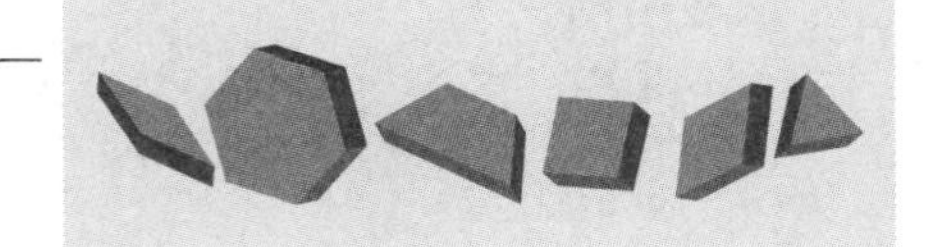

Pentagon

A five-sided polygon [see p. 99].

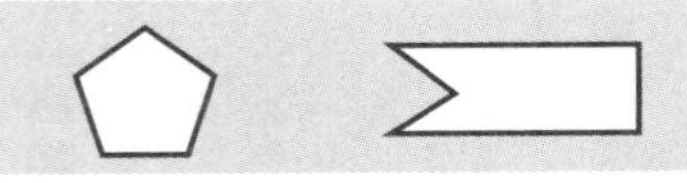

Pentomino

A polyomino made from 5 squares connected side-to-side. There are 12 pentominoes.

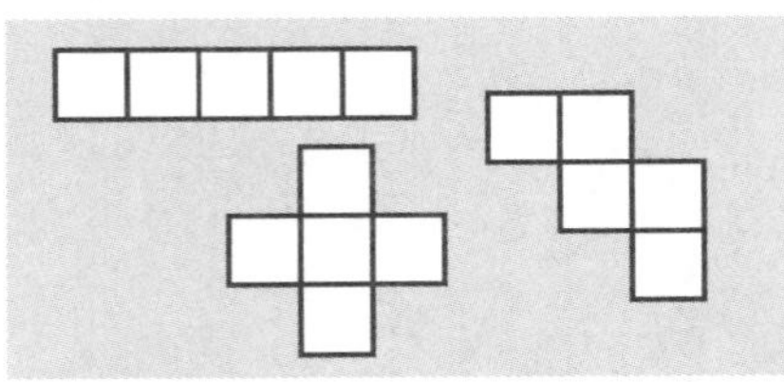

Percent (%)

A ratio with denominator 100; "percent" means "per hundred."

$$14\% = \frac{14}{100} = \frac{7}{50}$$

$$\frac{4}{5} = \frac{4 \times 20}{5 \times 20} = \frac{80}{100} = 80\%$$

Percentile

A point in a distribution below which a certain percent of data points fall.

In a distribution, 25% of the data points fall below P_{25}, the 25th percentile. P_{25} is also called the first quartile, Q_1.

Percent of change
(Percent of increase or decrease)

A percent that shows how much a quantity has increased or decreased compared to the original amount.

$$\text{Percent of change} = \frac{\text{Amount of increase (decrease)}}{\text{Original amount}}$$

The percent of change from 25 to 30 is $\frac{(30-25)}{25} = \frac{5}{25} = 20\%$.

List of Mathematical Terms

Perfect number

A counting number that is the sum of all its factors except itself.

6 is a perfect number because the factors of 6 are 1, 2, 3 and 6, and $1 + 2 + 3 = 6$.

Perfect square

A number that is the square of an integer.

$1 = 1^2$, $4 = 2^2$, $9 = 3^2$ and $16 = 4^2$ are perfect squares.

Perimeter

The distance around a plane figure, measured in linear units. [From the Greek *peri* meaning "around" and *metron* meaning "measure."]

The perimeter of this triangle is 14 cm.

6 cm
3 cm
5 cm

Permutation

An arrangement of objects in which order is important.

There are six possible permutations for the three letters A, B, C: ABC, ACB, BAC, CAB, BCA, CBA.

Perpendicular lines (⊥)

Two lines that intersect to form four right angles.

The two lines shown are perpendicular.
$\overleftrightarrow{AB} \perp \overleftrightarrow{CD}$

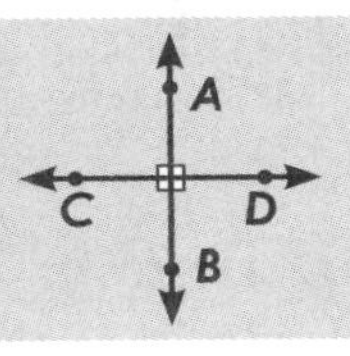

Pi (π)

The ratio of the circumference of a circle to its diameter. An irrational number that has a value of approximately 3.14.

Pictograph (Picture graph)

A graph that uses pictures or symbols to display data [see p. 95].

Ice creams sold at the cafeteria on Monday

A B C D E
Types of ice cream

Pie chart (Circle graph, Pie graph)

See Circle graph [see p. 95].

Pint (pt.)

A unit of capacity in the English system of measurement.

8 pints (pts.) = 4 quarts (qts.) = 1 gallon (gal.)

Place value

Assignment of value to the digits of a number based on their position within the number.

In the number 643.57, the 4 is in the tens place and has a value of 40; the 5 is in the tenths place and has a value of $\frac{5}{10}$.

List of Mathematical Terms

Plane

A flat surface that extends without end in all directions.

Plane of symmetry

A plane that cuts a 3-D object in half, such that each half is a mirror image of the other with the plane as the mirror [see p. 113].

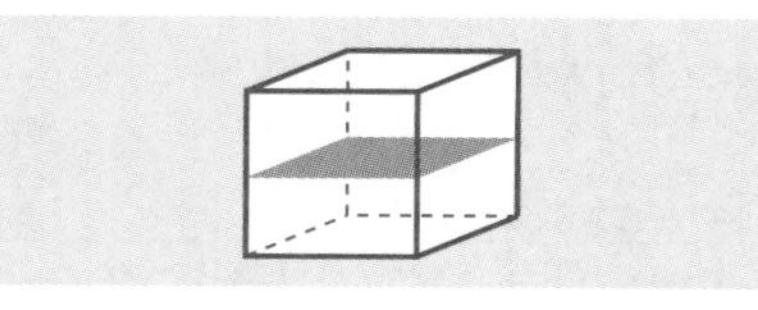

Platonic solids (Regular solids)

The five polyhedra in which all faces are congruent and all angles are congruent; i.e. the regular tetrahedron, hexahedron (cube), octahedron, dodecahedron and icosahedron. [Named after Plato the philosopher and mathematician in ancient Greece who stated that they represented the five elements of earth, wind, fire, water and the heavens or universe; see p. 111.]

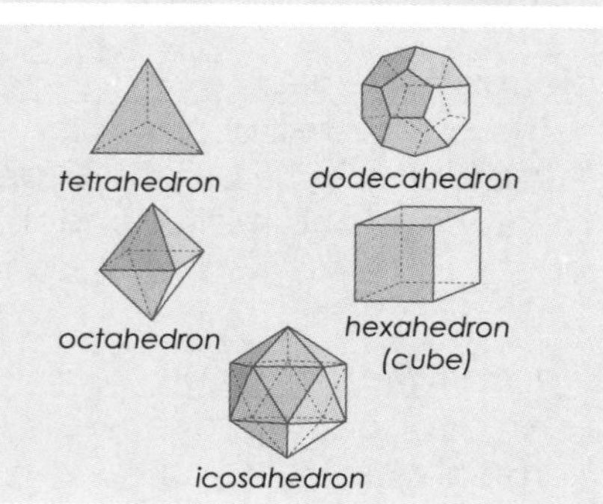

Plus (Add , +)

The expression 9 + 5 may be read as "nine and five," as "nine add five," or as "nine plus five."

p.m. (pm)

The time between noon (12:00 p.m.) and midnight (12:00 a.m.); Latin abbreviation for post meridiem.

We ate supper at 5:30 p.m.

Point

An undefined term in geometry that indicates a particular position.

Point symmetry

A planar symmetry in which a figure remains unchanged when rotated 180° about a point.

The figure has point symmetry about P.

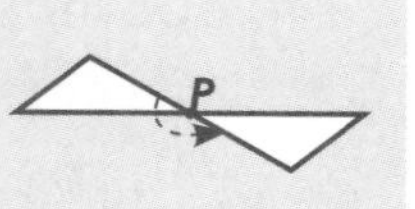

Polygon

Simple closed plane curve that is formed by three or more line segments. The intersection of two sides is called a vertex. [From the Greek *polus* meaning "many" and *gonia* meaning "angle."]

A triangle and a quadrilateral are polygons.

Polyhedra

Plural for polyhedron. See Polyhedron.

List of Mathematical Terms

Polyhedron

A three-dimensional shape that is formed by polygonal regions. The plural form is polyhedra. [From the Greek *polus* meaning "many" and *hedra* meaning "base"; see p. 109.]

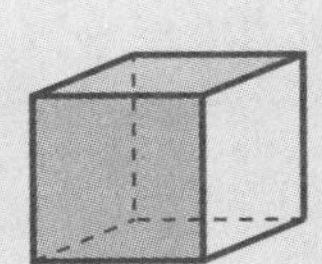

Polyomino

A plane shape of congruent squares, where each square is connected to at least one of the others by a common side, and the number of squares determines the name; with one example of four types shown here.

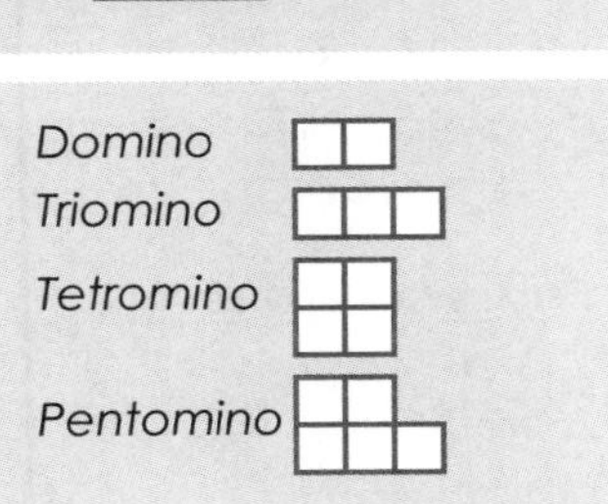

Polynomial

An algebraic expression with terms that consist of a coefficient and a variable raised to any positive integer power or the power zero. A monomial or a sum of monomial.

$3x$; $2x - 5y$;
$x^2 + 6x + 9$

Population

In statistics, the entire group to be considered.

- *All the girls in the school.*
- *All the counting numbers.*

Positive integers (I^+)

The integers greater that 0; the set of natural numbers, $I^+ = \{1, 2, 3, 4, 5, \ldots\} = N$.

2 and 57 are positive integers.

Pound (lb.)

A unit of measure of weight in the customary (English) system.

1 pound (lb.) = 16 ounces (oz.)

Power

1. A product formed from repeated multiplication by the same number or expression. A power consists of a base and an exponent.
2. The exponent *b* in the expression a^b.

- $2^3 = 2 \times 2 \times 2$ *is a power with base 2 and exponent 3. We read "2^3 is the third power of 2."*
- *In the expression 3^4, 4 is the power. we read "3 raised to the power 4."*

Preimage

The original figure prior to a transformation.

Image

m

m is the line of reflection.

List of Mathematical Terms

Prime factor

A prime number that will divide into a given counting number without remainder [see p. 77].

2, 5 and 7 are the prime factors of 70.

Prime factorization

A whole number written as the product of its prime factors.

The prime factorization of 48 is $2 \times 2 \times 2 \times 2 \times 3 = 2^4 \times 3$.

Prime number

A whole number greater than 1 whose only distinct factors are 1 and itself.

2, 3, 5 ,7, 11, 13, 17 and 19 are the first 8 prime numbers.

Principal

The amount of money that is deposited or borrowed.

If $100 is deposited in a bank account that pays 5% annual interest, the principal is $100.

Prism

A polyhedron formed by two congruent polygonal regions (bases) in parallel planes connected by parallelogram regions (lateral faces). A prism may be named by the shape of its bases.

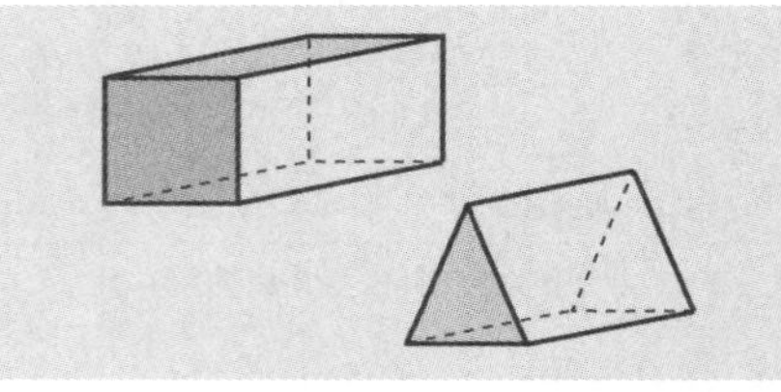

Probability

A measure of the likelihood that an event will occur; 0 ≤ probability of event ≤ 1.

In tossing a number cube, the probability of the event "a number less that 3 shows" is $\frac{2}{6} = \frac{1}{3}$.

Product

The result when two or more numbers are multiplied.

The product of 4 and 5 is 20; i.e. $4 \times 5 = 20$.

Proper fraction

A fraction, ignoring sign, where the numerator is less than the denominator.

$\frac{1}{4}$ $\frac{3}{8}$ $\frac{7}{10}$

Proportion

A mathematical sentence that equates two ratios.

$\frac{3}{4} = \frac{9}{12}$ *and 2:5 = 4:10 (read as "2 is to 5 as 4 is to 10")*

Protractor

An instrument for measuring angles.

Pyramid

A polyhedron formed by a polygonal region (base), a point (apex) not in the region, and the triangular regions connecting the point to the polygonal region. A pyramid can be named by the shape of its base [see p. 110].

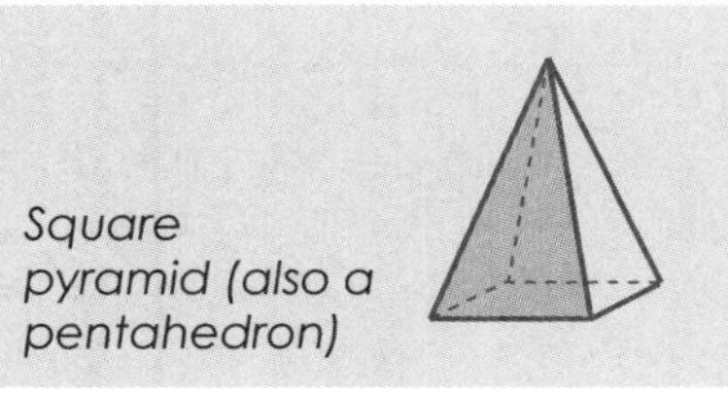

Square pyramid (also a pentahedron)

Pythagoras

The Greek mathematician who lived in the 6th century B.C. and whose name is given to the famous rule about right triangles; i.e. that the square on the hypotenuse is the sum of the squares on the other two sides; e.g. in the triangle shown, $3^2 + 4^2 = 5^2$.

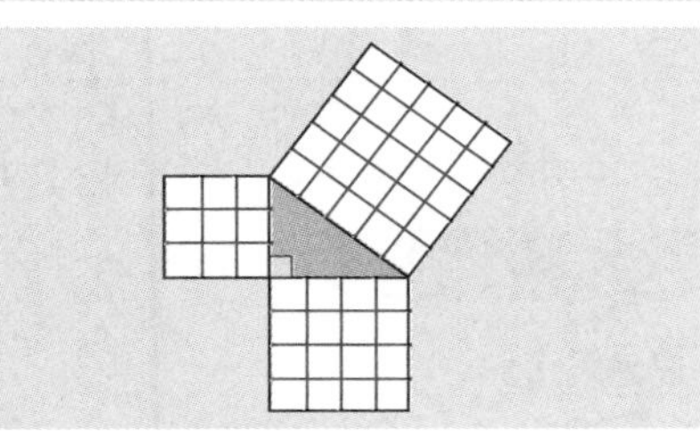

Pythagorean Theorem

For any right triangle, with legs of lengths a and b and hypotenuse of length c, the sum of the squares of the legs equals the square of the hypotenuse; or $a^2 + b^2 = c^2$.

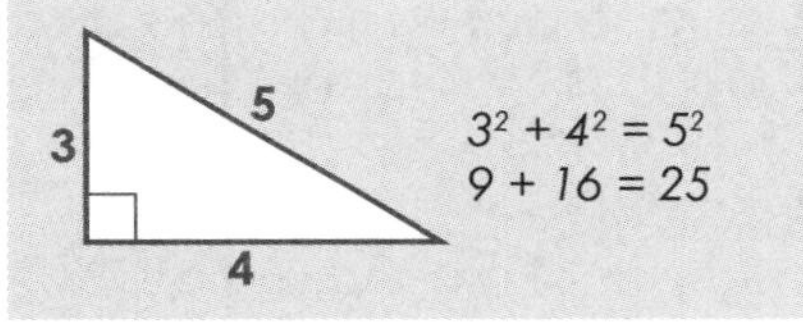

$3^2 + 4^2 = 5^2$
$9 + 16 = 25$

Pythagorean triple

A set of three positive integers a, b and c such that $a^2 + b^2 = c^2$.

5, 12, 13, is a Pythagorean triple.

Q

Quadrant

1. One of the four regions of the coordinate plane determined by the axes.
2. A quarter of a circular region.

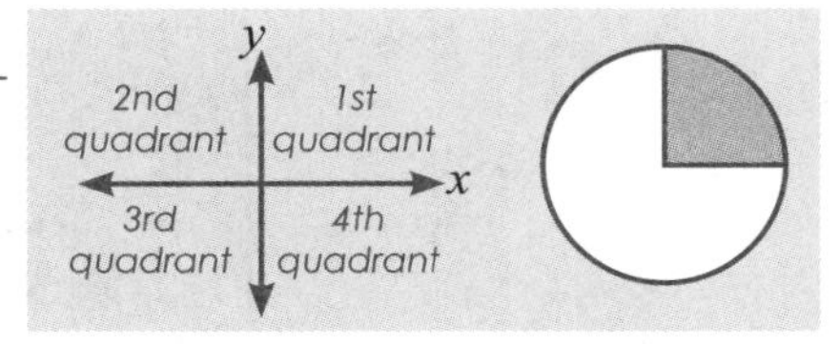

Quadratic equation

An equation of the form $ax^2 + bx + c = 0$, where $a \neq 0$. The expression $ax^2 + bx + c$ is called a quadratic expression (quadratic polynomial).

$3x^2 - 1 = 0$ $\quad x^2 + 4x = 0$
$x^2 + 5x + 6 = 0$

Quadrilateral

Polygon that has four sides. [From the Latin *quadri* meaning "four" and *latus* meaning "side"; see pp. 104-105.]

Quart (qt.)

A unit of capacity in the customary (English) system of measurement.

1 gallon (gal.) = 4 quarts (qts.) = 8 pints (pts.)

List of Mathematical Terms

Quartile

Point in a data set below which 25%, 50%, or 75% of the data lie. The quartiles are Q_1 (first quartile or lower quartile), Q_2 (second quartile or median), and Q_3 (third quartile or upper quartile).

In terms of percentiles, $Q_1 = P_{25}$, $Q_2 = P_{50}$ and $Q_3 = P_{75}$.

Quotient

The result of a division. [From the Latin *quotiens* meaning "how many times."]

When 24 is divided by 8, the quotient is 3.

R

Radian

The size of an angle at the center of a circle formed by two radii and an arc that is the length of the radius.

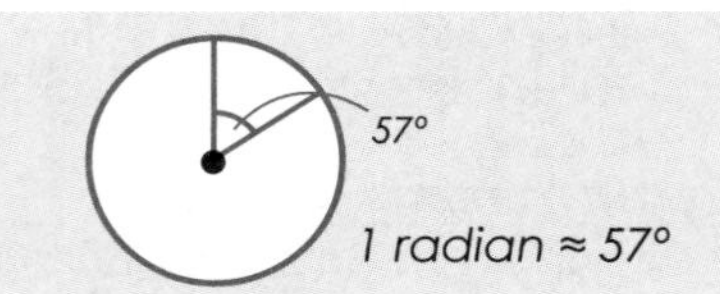

1 radian ≈ 57°

Radical $\sqrt{\ }$

A symbol used to denote the positive square root of a number. The symbol ($\sqrt[n]{\ }$) denotes the *n*th root of a number.

$\sqrt{9} = 3$, the positive square root of 9 is 3
$\sqrt[3]{8} = 2$, the cube root of 8 is 2
$\sqrt[4]{81} = 3$

Radius (of a circle)

A line segment from the center of a circle to any point on the circle.

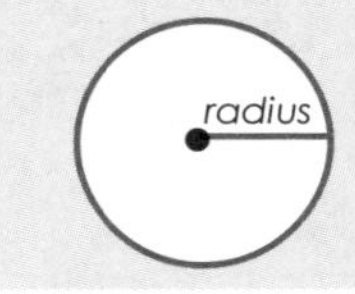

Random sample

A sample chosen from a population in such a way that all members have an equal chance of being selected.

Range

In statistics, the difference between the greatest and least values in a set of data.

The range of the data set 15, 23, 37, 42, 58, 76 is $76 - 15 = 61$.

Range (of a function)

The set of all output values for a function.

Input	*1*	*2*	*3*	*4*
Output	*3*	*6*	*9*	*12*

The range is {3, 6, 9, 12}.

Rate

A ratio of two quantities measured in different units.

65 mi/hr
60km/hr

Ratio
A comparison of one number to another number.

The ratio of 3 to 4 can be expressed as $\frac{3}{4}$ or as 3:4.

Rational numbers (Q)
Numbers that can be written as $\frac{a}{b}$, where *a* and *b* are integers and $b \neq 0$.

Real numbers that can be expressed as terminating or repeating decimals.

$$3 = \frac{3}{1},\ 3.7 = \frac{37}{10},\ 2\frac{1}{4} = \frac{9}{4},\ \frac{-2}{3},\ -4\frac{1}{2}$$

$$.3 = .3333\ldots = \frac{1}{3}$$

$$.25 = \frac{1}{4}$$

$$.45 = \frac{5}{11}$$

Ray ($\overrightarrow{AB}$)
A part of a line that has one endpoint and extends without end in one direction.

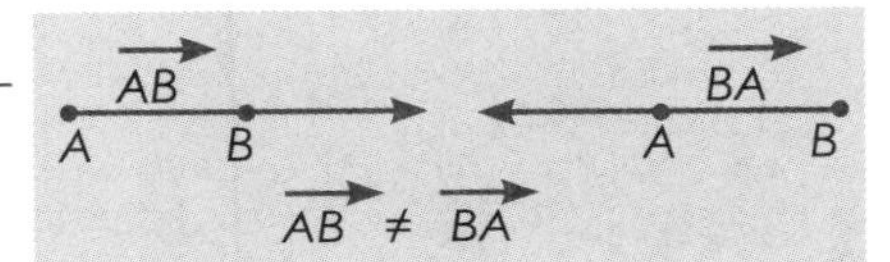

Real numbers
The set of all rational numbers and irrational numbers; $R = Q \cup Q'$.

The set of numbers that can be represented on a number line.

$$0,\ \frac{3}{4},\ \frac{-5}{8},\ 7,\ -2,\ \varpi, \sqrt{2},\ 5 + 4\sqrt{3}$$

Reciprocal (Multiplicative inverse)
One of two numbers whose product is 1.

$\frac{2}{3}$ is the reciprocal of $\frac{3}{2}$ and vice versa; the reciprocal of any number n is $\frac{1}{n}$.

Rectangle
A parallelogram with four right angles.

Rectangular numbers (Composite numbers)
Numbers that can be represented by objects in a rectangular array with more than one row and more than one column.

Square numbers greater than 1 are rectangular numbers.

Rectangular numbers are composite numbers.

○○○ ☆☆☆☆☆
○○○ ☆☆☆☆☆
☆☆☆☆☆

3 is not a rectangular number since it can only be represented by 1 row: ☆☆☆

Rectangular prism
A prism with rectangular bases.

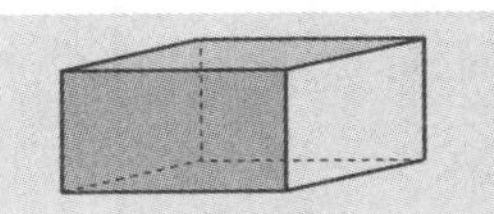

Recurring decimal
See Repeating decimal.

List of Mathematical Terms

Reduction
See Contraction.

Reflection (Flip)
A transformation that reflects a figure (preimage) in (or about) a line, called the line of reflection, creating a mirror image of the figure.

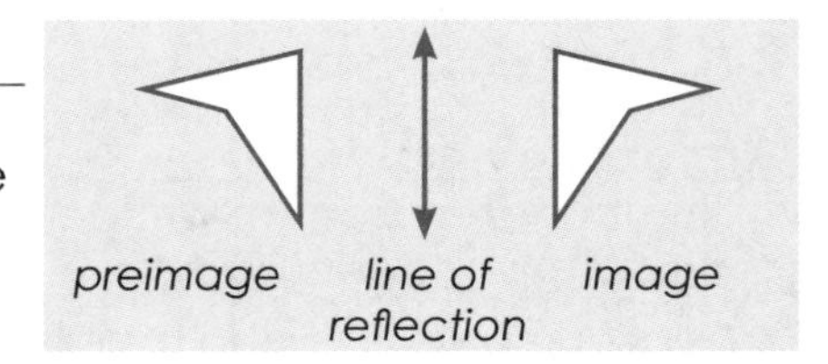

Reflective symmetry
See Line symmetry.

Reflex angle
An angle with measure greater than 180° and less than 360°.

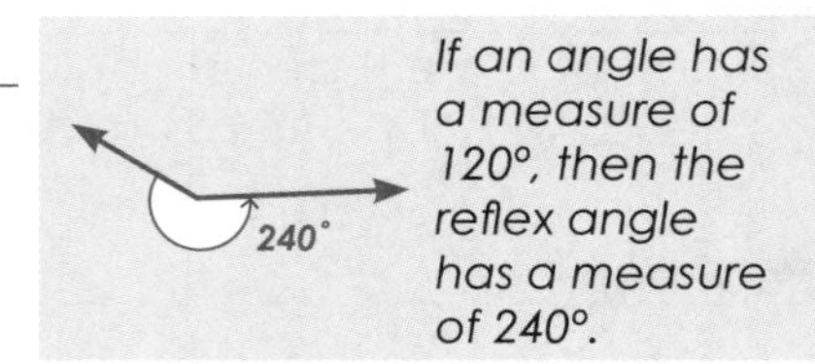

Region
Part of a plane or part of space. An interior region is the set of all points on the inside of a simple closed curve and an exterior region is the set of all points on the outside of a simple closed curve.

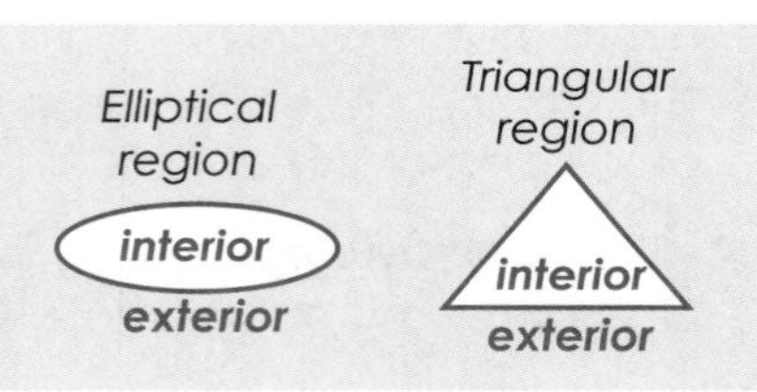

Regular polygon
A polygon with congruent sides and congruent interior angles.

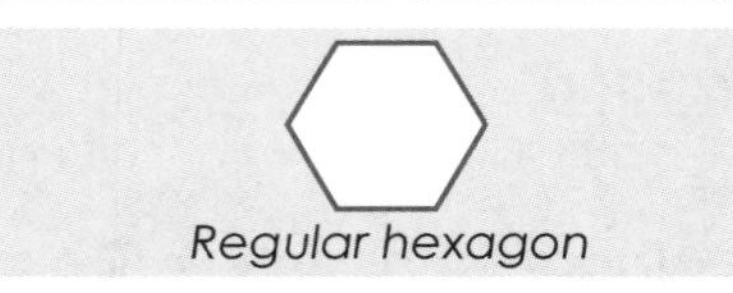

Regular polyhedron (Regular solid, Platonic solid)
A solid with congruent polygonal faces and congruent interior angles. The five Platonic solids are the regular polyhedra.

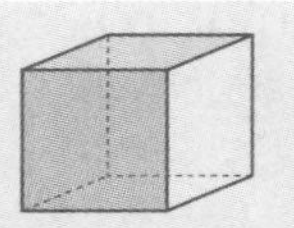
Cube or hexahedron

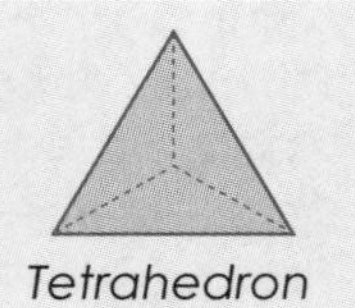
Tetrahedron

Regular tessellation
A tessellation made from only one type of regular polygon.

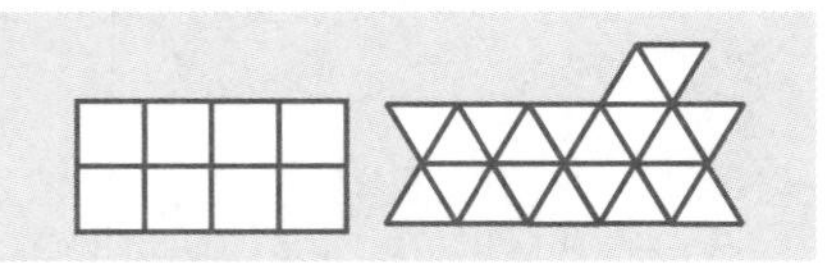

Relation
A set of ordered pairs.

{(3,5), (4,9), (0,–2), (5,1), (3,4), (4,10)} is a relation.

Relatively prime numbers

Two or more counting numbers for which the greatest common factor is 1. This concept can be extended to integers.

9 and 16 are relatively prime since the GCF is 1.
12 and 15 are not relatively prime since the GCF is 3.

Remainder

The whole number left over after a division of whole numbers.

When dividing 10 by 3 the remainder is 1.

Repeating (recurring) decimal

A decimal where one or more digits continually repeat.

$0.345345 = 0.\overline{345}$

Revolution

One complete turn through 360 degrees.

Rhombus

A parallelogram with four congruent sides. [Often incorrectly called a diamond; see p. 104.]

Right angle

An angle that has a measure of exactly 90°.

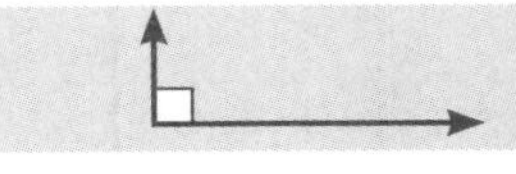

Right triangle

A triangle with one right angle.

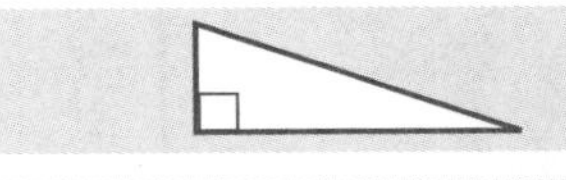

Rise

The vertical change between two points on a line.

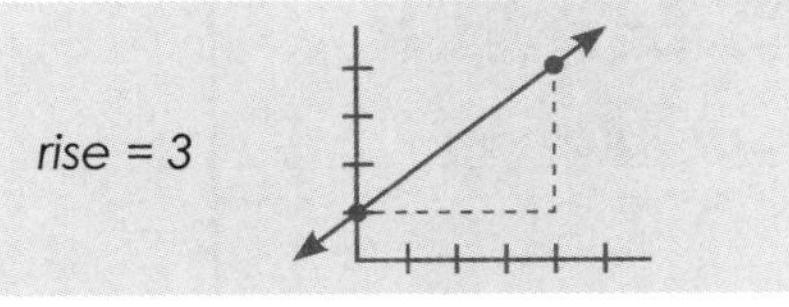

Roman numerals

Ancient system of numeration where numbers are represented by letters of the Roman alphabet; i.e. the numerals are formed from a combination of the symbols I (1), V (5), X (10), L (50), C (100), D (500), and M (1000); [see p. 62].

CLXIV = 164

Root

The base of a power.

In $2^3 = 8$, the cube root of 8 is 2, shown as $\sqrt[3]{8} = 2$.

Root (of an equation)

A number that satisfies the equation; also called the solution of the equation.

$y = 5$ is the root or solution of the equation, $3y + 2 = 17$.

Rotation (Turn)

A transformation that turns a figure in a given direction through a given angle (angle of rotation) about a fixed point (center of rotation).

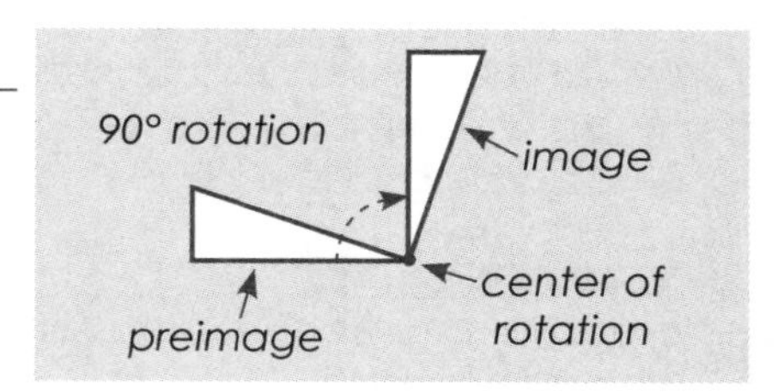

Rotational symmetry

A figure has rotational symmetry if a turn of 180° or less produces an image that fits exactly on the original figure.

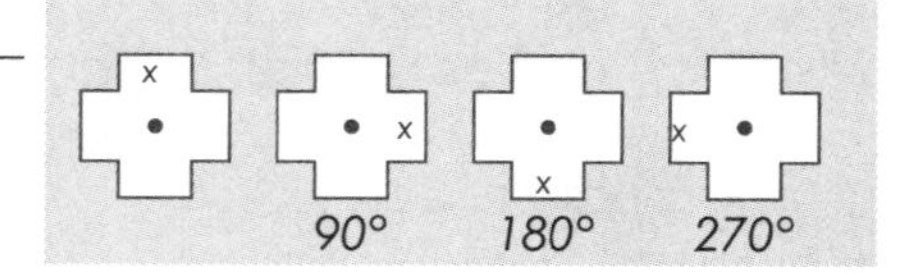

Rounding

Approximating a number to the nearest specified decimal place.

4873 rounded to the nearest:
ten is 4870
hundred is 4900
thousand is 5000

Row

An arrangement of objects in a horizontal line. See Array.

Run

The horizontal change between two points on a line.

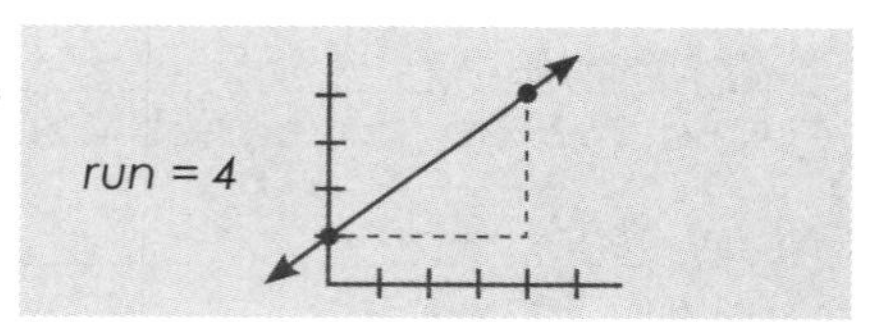

S

Sample

A part of a population.

Sample space

The set of all possible outcomes for an experiment.

In tossing a number cube, the sample space, S, is S = {1, 2, 3, 4, 5, 6}.

Scale

In a scale drawing, the relationship between the drawing's dimensions and the actual dimensions.

If the scale on a map is 1:100,000 then 1 cm on the map represents 100,000 cm = 1 km at the actual location.

Scale drawing

A diagram of an object in which the dimensions are in proportion to the actual dimensions of the object.

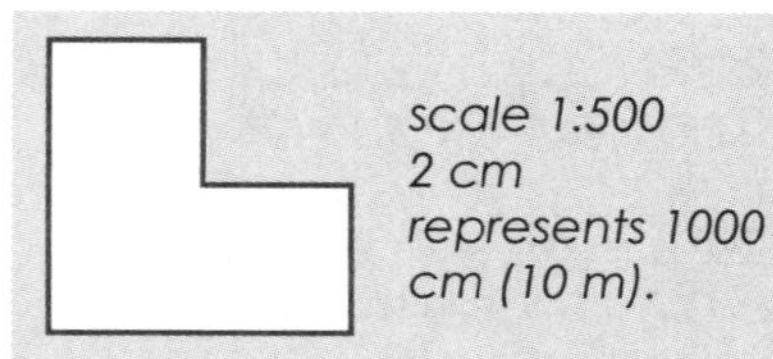

Scale factor

The ratio of corresponding side lengths of the image and its preimage under a dilation.

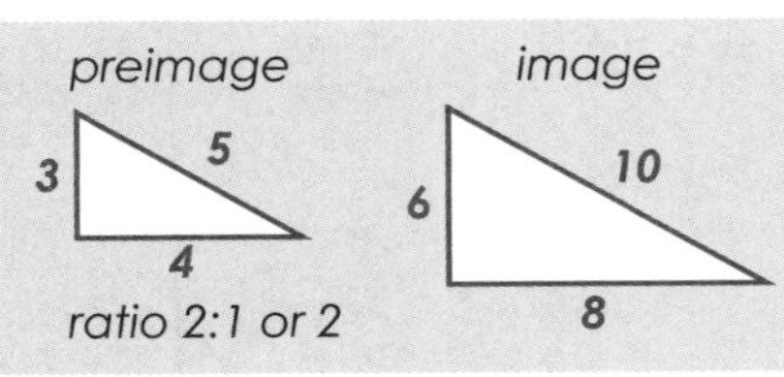

Scale model

A model of an object in which the dimensions are in proportion to the actual dimensions of the object.

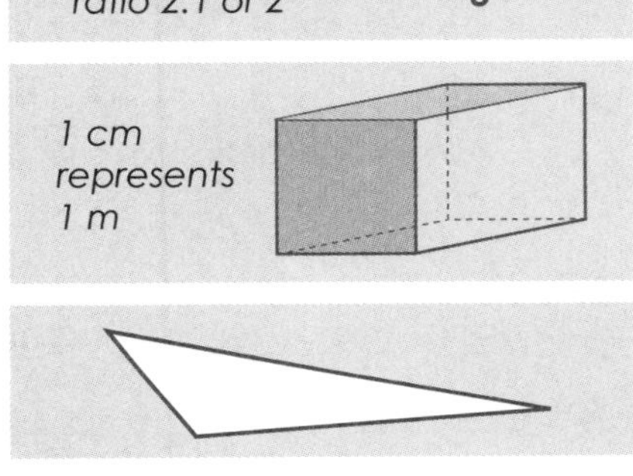

Scalene triangle

A triangle with three sides of different lengths. [From the Greek *skalenos* meaning "uneven."]

Scales

Instruments used for comparing or measuring masses; the main types being a beam balance, spring-based scales and electronic scales.

Scatterplot (Scattergram)

The graph of a set of data pairs (x, y), which is a set of points in the coordinate plane.

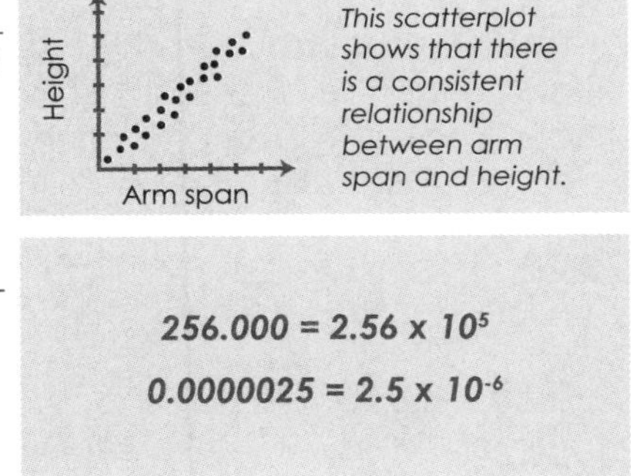

This scatterplot shows that there is a consistent relationship between arm span and height.

Scientific notation

A method for expressing a number as a product $c \times 10^n$, where c is greater than or equal to 1 and less than 10, and n is an integer.

256.000 = 2.56×10^5

0.0000025 = 2.5×10^{-6}

Score

A value in statistics, and also a name for 20. [From Old Norse *skor* meaning a "notch" or "tally mark," and with marks grouped in 20s this led to the use of it as a name for 20.]

Second (s)

1. A unit of time or angle measure; 60 seconds = 1 minute.
2. Position number two in order.

Sector

Portion of a circular region bounded by two radii and an arc [see p. 107].

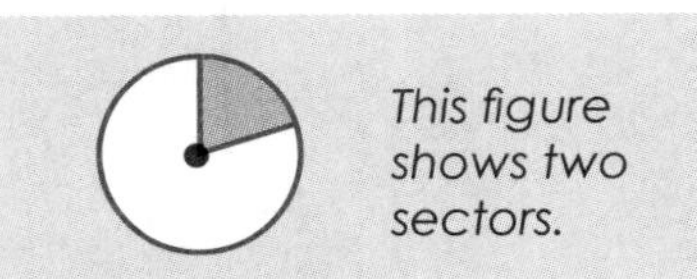

This figure shows two sectors.

List of Mathematical Terms

Segment

See Line segment.

Semicircle

Half a circle; any diameter divides a circle into two semicircles.

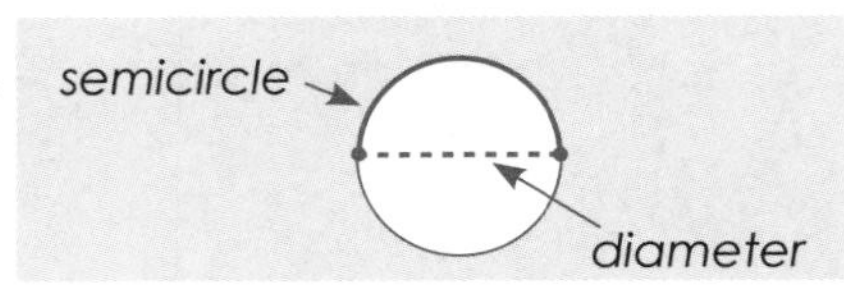

Sequence

An ordered arrangement of objects, figures, or numbers.

3, 6, 9, 12, 15, … is the sequence of multiples of 3.

Series

A sum of a sequence of numbers.

$1 + 2 + 3 + 4 + \ldots + n$

Set

A well-defined collection of objects or numbers.

The set of square numbers less than 10 is {1, 4, 9}.

Set theory

The branch of mathematics that studies properties of sets.

SI

The international system of measurement units derived from the metric system and using the base units of meter, kilogram, second, ampere, candela, Kelvin and mole. [SI is the universal symbol for the French *Système Internationale d'Unités*; see pp. 118-199.]

Side

A line segment forming part of a polygon.

An octagon has eight sides.

Significant figure

The digit in a numeral that indicates the size of the number to a certain degree of accuracy.

In 359 the 3 is the most significant figure.

Similar

Having the same shape but not necessarily the same size.

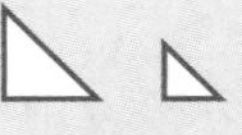

similar triangles

similar shapes

Simple closed curve

A curve that does not cross itself and starts and stops at the same point.

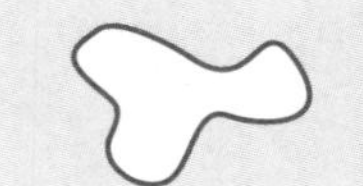

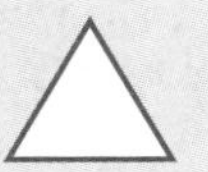

Simple curve

A curve that does not cross itself, except possibly at the starting and stopping points.

simple

not simple

Simple interest

Interest that is earned or paid only on the principal. The simple interest, I, is the product of the principal, P, the annual interest rate, r, written as a decimal, and the time, t, in years. $I = Prt$.

If \$600 is deposited in a savings account for 1 year with a (simple) interest rate of 5%, then the interest is I = \$600 x .05 x 1 = \$30.

Simplest form (Simplest terms)

A fraction is in simplest form if its numerator and denominator have a greatest common factor of 1.

The simplest form of the fraction $\frac{9}{12}$ is $\frac{3}{4}$.

Sine

For any acute angle of a right triangle, the ratio of the length of the opposite leg to the length of the hypotenuse.

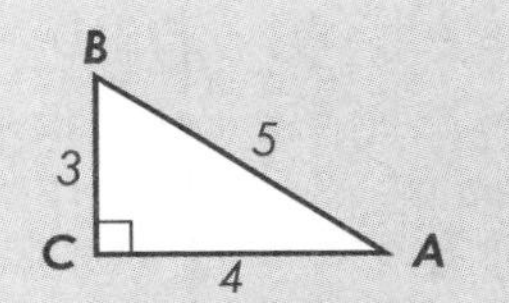

$\sin A = \frac{3}{5}$

$\sin B = \frac{4}{5}$

Size transformation (Dilation)

A transformation that stretches or shrinks a figure (expansion or contraction).

Skew lines

Lines in different planes that do not intersect.

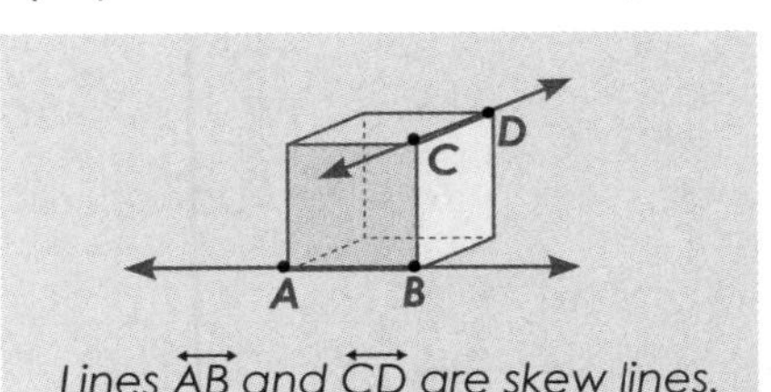

Lines $\overleftrightarrow{AB}$ and $\overleftrightarrow{CD}$ are skew lines.

Slant height

The height of a triangular face that is not the base of a pyramid.

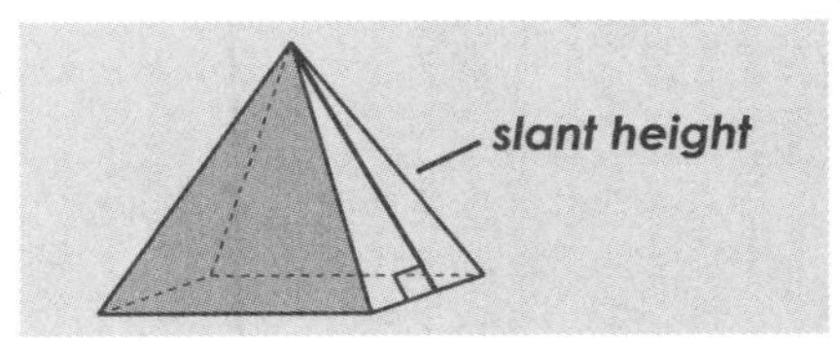

Slide

See Translation.

Slope (Gradient)

The ratio of the rise (vertical change) to the run (horizontal change) between any points on a nonvertical line. (The slope of a vertical line is undefined.)

(4,4)

(0,1)

3

4

$\text{slope} = \frac{\text{rise}}{\text{run}} = \frac{4-1}{4-0} = \frac{3}{4}$

List of Mathematical Terms

Slope – Intercept form

The form of a linear equation $y = mx + b$, where m is the slope and b is the y-intercept.

$y = 2x + 5$, $y = -\frac{1}{3}x - 7$

Solid

A 3-dimensional shape. [In the classic definition, a solid must contain its interior.]

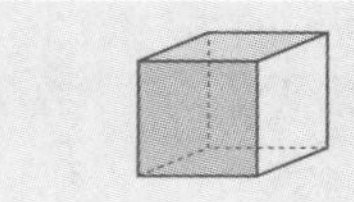

Solution (of an equation or inequality in one variable)

A number that, when substituted for the variable in the equation or inequality, makes the statement true.

The solution of the equation $n + 4 = 9$ is 5.
The solution of the inequality $x + 3 < 9$ is < 6.

Solution (of a linear equation or inequality)

An ordered pair (x, y) that makes the statement true when the values of x and y are substituted into the equation or inequality.

(2, 7) is a solution of $y = 3x + 1$.
(3, 5) is a solution of the inequality $y \geq 2x - 4$.

Solve (an equation or inequality)

To find all the solutions of an equation or inequality by finding all the values of the variable or variables that make the statement true.

Speed

Distance travelled per unit of time.

60 km/h

Sphere

A surface formed by all the points in space that are the same distance from a fixed point, called the center.

- *a tennis ball is a model of a sphere*
- *the Earth is almost spherical*

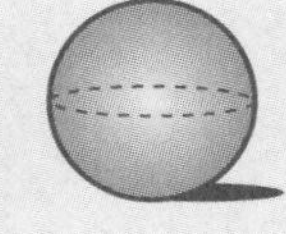

Square

1. A rectangle with four congruent sides. A parallelogram with four right angles and four sides of equal length.
2. The product of a number and itself.

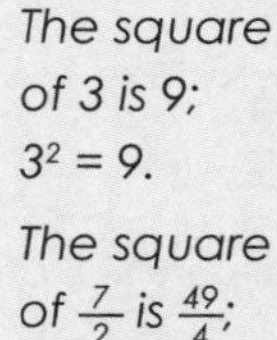

The square of 3 is 9;
$3^2 = 9$.

The square of $\frac{7}{2}$ is $\frac{49}{4}$;
$(\frac{7}{2})^2 = \frac{49}{4}$.

Square number

Any counting number obtained by multiplying a number by itself. A square number can be represented by a square array of dots.

$1 = 1^2$, $4 = 2^2$, $9 = 3^2$ are square numbers.

Square root ($\sqrt{\ }$)

A square root of a number *n* is a number *m* which, when multiplied by itself, equals *n*.

The opposite of squaring a number.

The square root of 25 is 5 ($\sqrt{25} = 5$), since $5^2 = 25$.

Square unit

A unit of area equivalent to that contained in a square region with side length 1 unit.

In the metric system, square centimeters (cm^2), square meters (m^2), and square kilometers (km^2) are popular choices.

In the customary (English) system, square inches (in^2), square feet (ft^2), and square miles (mi^2) are popular choices.

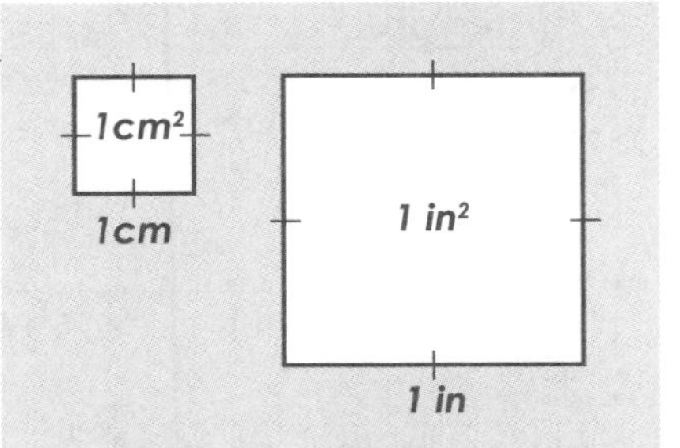

Standard deviation

A statistical measure of the spread, or dispersion, of a data set; describes how far, on average, the points are from the mean.

Standard form

A polynomial written with the exponents of the variable decreasing from left to right.

$4x^3 - 6x^2 + 7x - 5$ is in standard form.

Stem

All the digits except the last digit on the right of a number displayed in a stem-and-leaf plot.

Stem-and-leaf plot

A data display that organizes data points by separating each into a leaf (last digit) and a stem (remaining digits).

Stem	Leaves
1	9 6 9
2	1 7 6 4 8
3	4 5 2

A stem-and-leaf plot for data 21, 19, 34, 16, 27, 35, 19, 26, 24, 28, 32

Straight angle

An angle whose measure is exactly 180°.

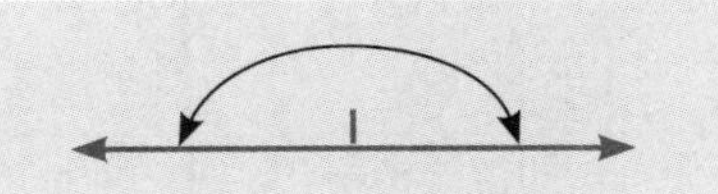

Subset ($\subseteq$ or $\subset$)

A set within a set.

If $A = \{1, 2, 4\}$ and $B = \{1, 2, 3, 4, 5\}$ then A is a subset of B.

$A \subseteq B$

List of Mathematical Terms

Subtraction (–)

A numerical operation involving "taking away" or "finding the difference" or "finding the complement."

$10 - 6 = ?$

$6 + ? = 10$

Subtraction property of equality

Subtracting the same number from each side of an equation produces an equivalent equation.

If $x + 6 = 10$, then
$x + 6 - 6 = 10 - 6 = 4$.

If $x + a = b$, then
$x + a - a = b - a$.

Sum

The result of adding two or more numbers. See Addition.

The sum of 8 and 9 is 17, since $8 + 9 = 17$.

Supplementary angles

Two angles whose measures have a sum of 180°.

135° 45°
A B C D

Two right angles.

$\angle ABC$ and $\angle DBC$ are supplementary.

Surface

The outer boundary or one of the boundary regions of a 3-dimensional shape.

The top and bottom surfaces are circular regions while the remainder is a curved surface.

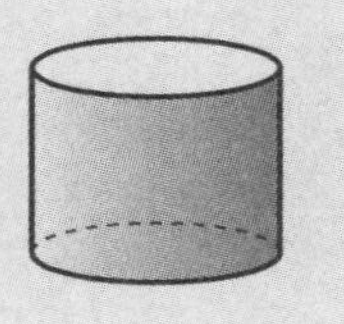

Surface area

The sum of the areas of the outside surfaces or faces of a 3-dimensional shape.

surface area = 2(1) + 4(3)
= 14 square units

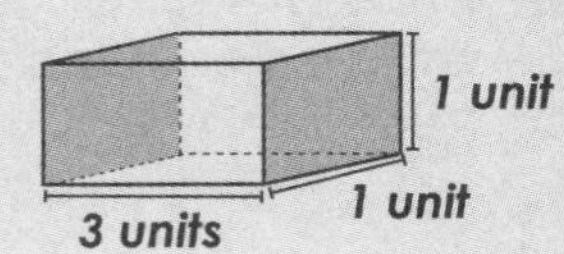

Survey

A method of gathering information about a population.

Symbol

A letter, numeral, or mark that represents something.

8, +, –, x, ÷, g, =, %, >

Symmetry

A transformation that preserves the appearance of a figure. See Point symmetry, Line symmetry and Rotational symmetry.

The triangle is symmetrical about the vertical line, an example of line symmetry.

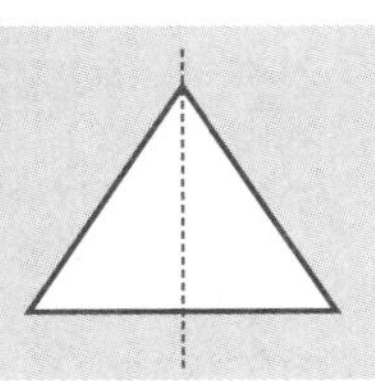

T

Table

A means of organizing data in rows and columns for a particular purpose.

	Red socks	Blue socks
Jen	2	6
Ryan	6	8

Tally marks

Marks made to record items or events, usually grouped in fives by a diagonal stroke [see p. 89].

The diagram shows a tally of 14 objects.

Tangent

1. For any acute angle of a right triangle, the ratio of the length of the opposite leg to the length of the adjacent leg.

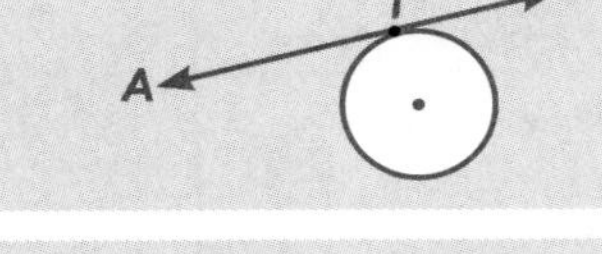

2. A line that intersects a curve at exactly one point. From the Latin *tangere*, meaning "to touch."

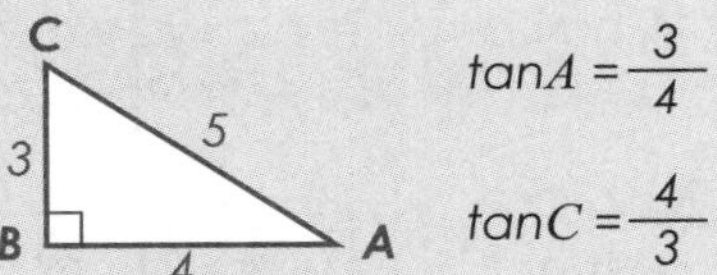

Tangram

An ancient Chinese square puzzle consisting of seven pieces that can be rearranged to make various shapes.

Terminating decimal

A decimal that has a final digit.

$\frac{3}{8} = 3 \div 8 = 0.375$, which is a terminating decimal.

Terms

The parts of an expression that are added or subtracted.

The terms of $4x^2 + 5x + 2$ are $4x^2$, $5x$, and 2.

Tessellation (Tiling)

A covering of a plane with congruent copies of the same pattern so that there are no gaps or overlaps.

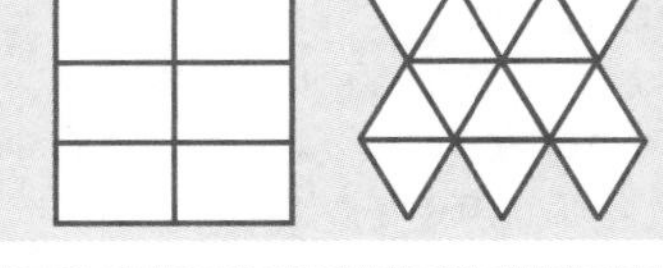

Tetrahedron (Triangular pyramid)

A polyhedron with four triangular faces. A regular tetrahedron is one of the five Platonic solids.

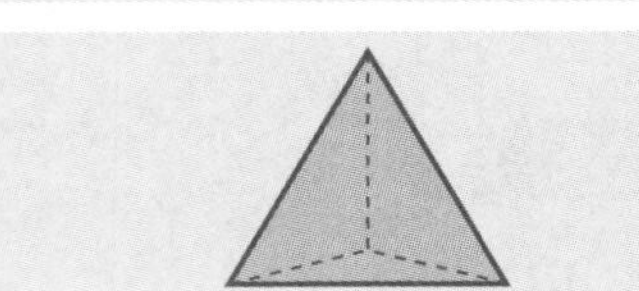

Theoretical probability

A probability based on the outcomes of an experiement under ideal conditions. In the case of equally likely outcomes, the theoretical probability of an event is given by:

$$P\ (\text{Event}) = \frac{\text{Number of favorable outcomes}}{\text{Total number of outcomes}}$$

A bag of 10 marbles contains 4 blue and 6 white marbles. The theoretical probability of randomly choosing a blue marble is $\frac{4}{10} = 0.4$.

Three-dimensional (3-D)

See Dimension.

Tonne (t)

A unit of mass equivalent to 1000 kg. [Not to be confused with ton which is an English measure of 2000 pounds.]

Topology

A branch of geometry concerned with the properties of figures that are independent of measurement; i.e. the properties remain the same even when the figure is stretched or distorted. [Topology is often called "rubber sheet geometry."]

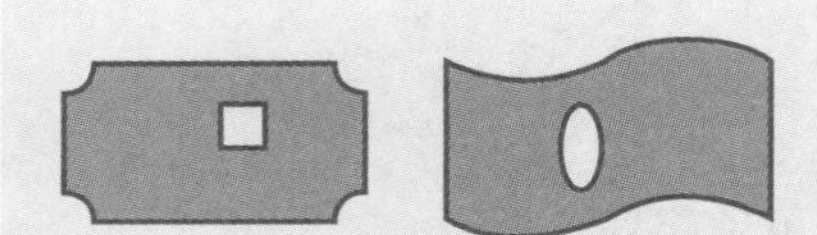

The two figures shown are topologically the same as they both have one hole.

Transformation

A movement of points in the plane which may change the shape, size, or position of the figure [see p. 114].

Translation (Slide)

A transformation that moves each point of a figure the same distance in the same direction.

Figure B is the image of Figure A under a translation.

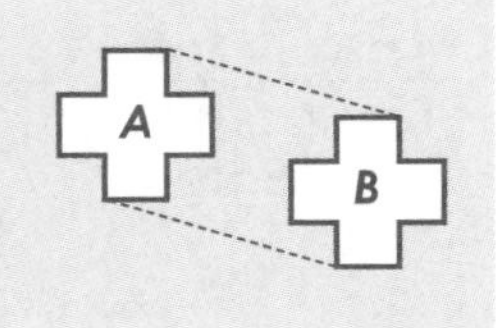

Transversal

A line which intersects two or more (often parallel) lines.

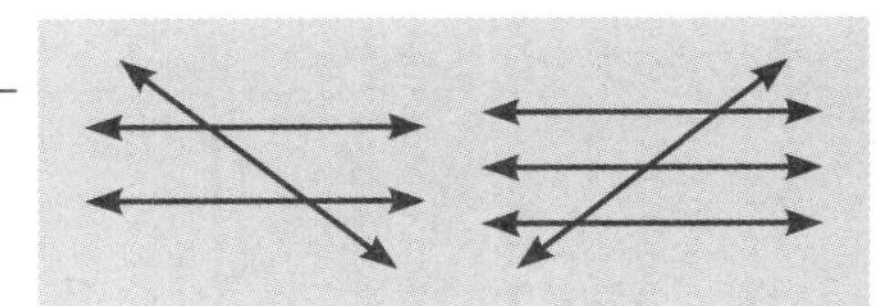

Trapezoid

1. A quadrilateral with at least one pair of parallel sides. [From the Greek *trapeza* meaning "table."]
2. Older definition: A quadrilateral with exactly one pair of parallel sides.

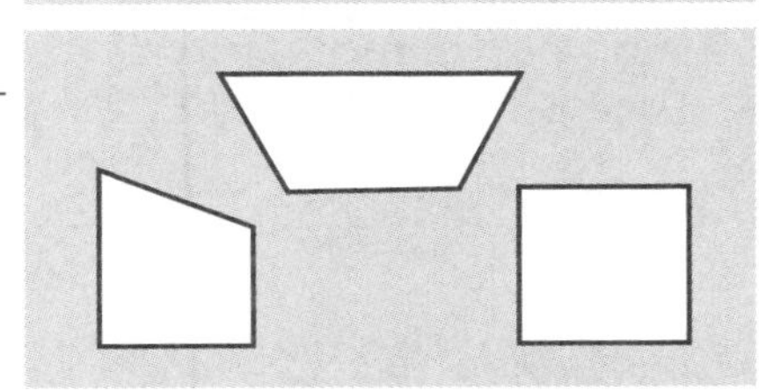

Traversable network

A network with a path that includes each edge (arc) exactly once. See Euler circuit.

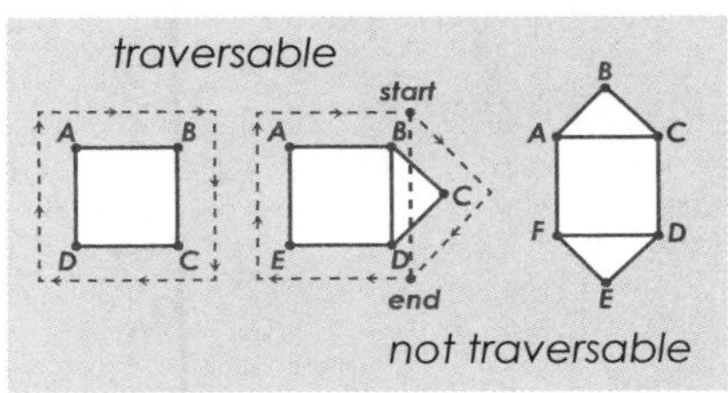

Tree diagram

A branching diagram that shows all the possible choices or outcomes of a process carried out in several stages.

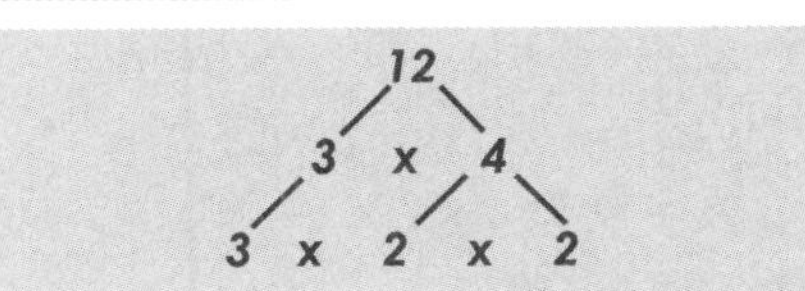

Triangle

A polygon with three sides.

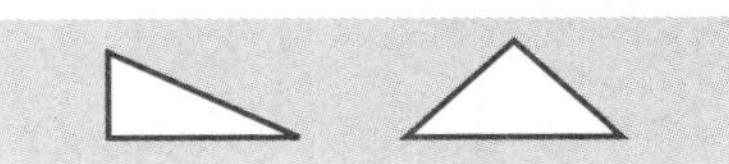

Triangle inequality

The sum of the lengths of any two sides of a triangle is greater than the length of the third side.

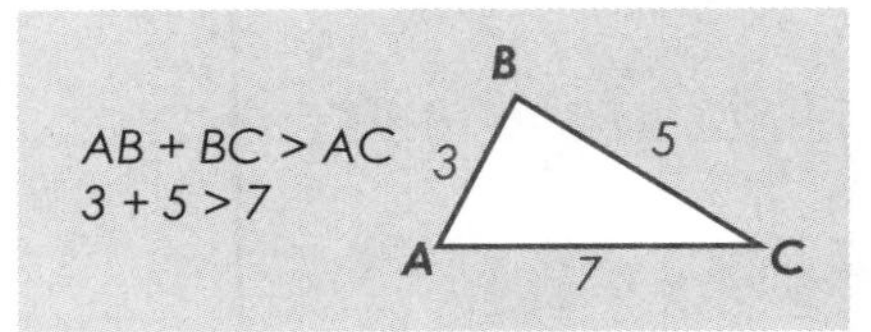

Triangular number

A number that can be represented by objects or symbols arranged in the shape of a triangle.

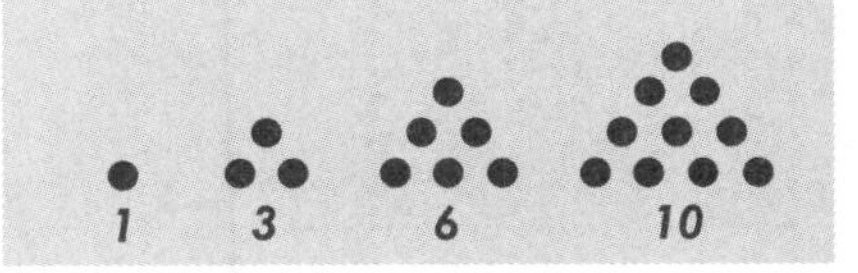

Triangular prism

A prism with triangular bases.

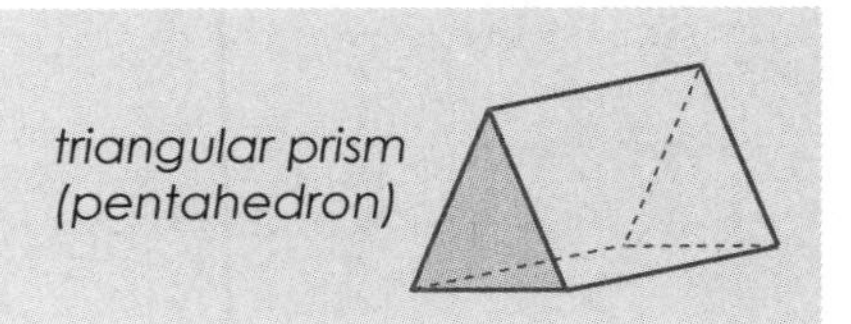

List of Mathematical Terms

Trigonometric ratio

A ratio of the lengths of two sides of a right triangle. See Sine, Cosine and Tangent.

Trigonometry

A branch of mathematics concerned with the relationships between the angles and the sides of triangles.

Trillion

The number 1,000,000,000,000 = 10^{12} [see p. 73].

Trinomial

A polynomial with three terms.

$4x^2 - 5x + 7$

Trundle wheel

A wheel used to measure distance.

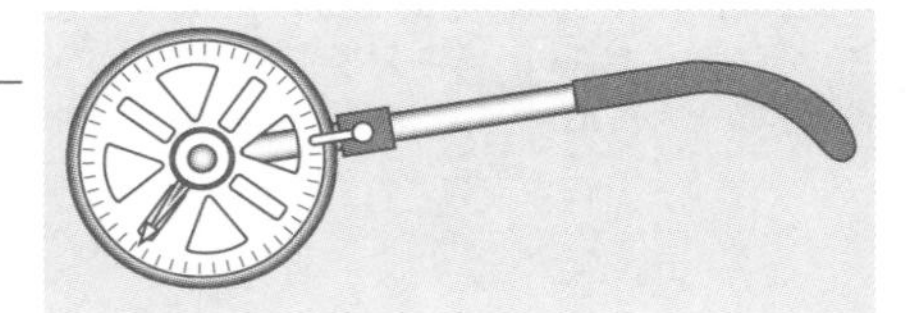

Turn symmetry

See Rotational symmetry.

Two-dimensional (2-D)

See Dimension.

U

Unequal

See Not equal.

Unfavorable outcome

An outcome that is not favorable.

Union (∪)

A set consisting of all the elements in two or more sets.

If A = {1, 3, 5} and B = {1, 4, 6,}, then A ∪ B = {1, 3, 4, 5, 6}.

Unit

1. A standard for measurement.
2. Another name for the number 1.

Some metric units for length are mm, m and km.

Units place

The ones place in the base ten numeration system.

In 259, the digit 9 is in the units place.

Unit rate
A rate that has a denominator of one.

$6.25 per (1) hour
60 miles per (1) hour

Universal set (U)
The set of all members being considered.

If the odd and even numbers are being considered then the universal set for this case could be the set of all whole numbers.

Upper extreme (value)
The greatest value in a data set.

In the data set {2, 5.5, 3.2, 1, 7}, 7 is the upper extreme value.

Upper quartile (Third quartile, Q_3)
The median of the upper half of a data set. As a percentile, P_{75}.

V

Variable
A symbol, usually a letter, that represents one or more numbers.

In the equation $x + y = 10$, x and y are variables while 10 is a constant.

Variable expression
See Algebraic expression.

Velocity
Speed in a particular direction.

Venn diagram
A diagram that uses shapes to represent sets and their relationships. [Named after John Venn who developed the method in the 1890s; see p. 89.]

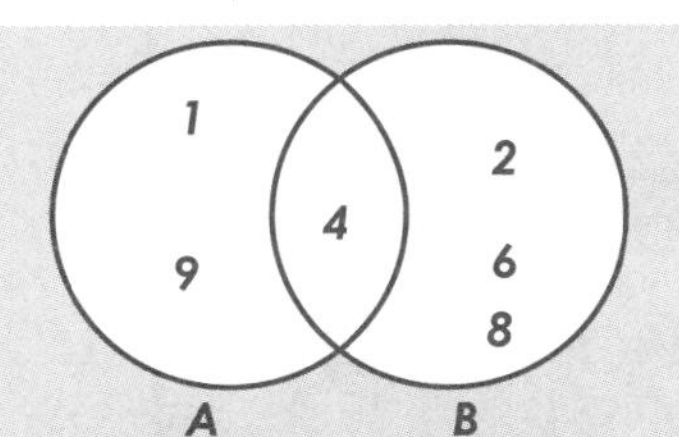

Venn diagram showing the relationship between sets A = {1, 4, 9} and B = {2, 4, 6, 8}.

Verbal model
A word equation that represents a real-world situation.

Distance is the product of rate and time ($d = r \times t$).

Vertex
A point where two rays meet, where sides of a polygon meet, or where edges of a polyhedron meet.

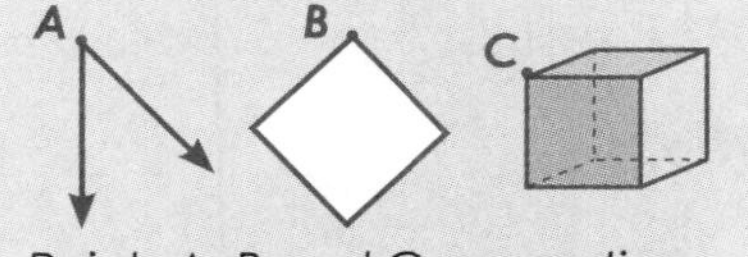

Points A, B and C are vertices.

List of Mathematical Terms

Vertical
The direction of a plumb line; i.e. upright and perpendicular to the horizontal.

Vertical angles
A pair of opposite angles formed by intersecting lines.
Vertical angles are congruent.

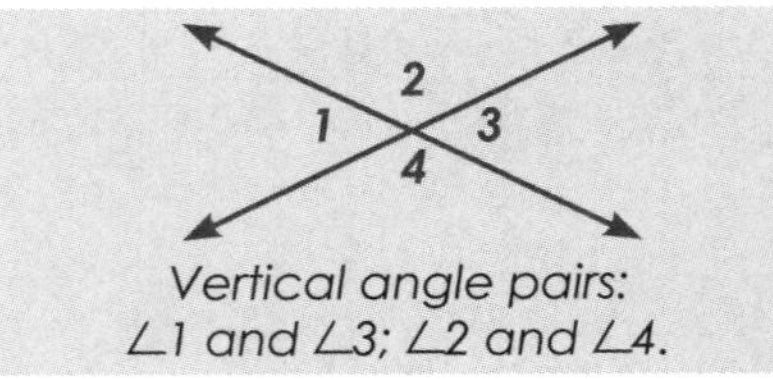

Vertical angle pairs:
$\angle 1$ and $\angle 3$; $\angle 2$ and $\angle 4$.

Vertical line test
If a vertical line intersects a graph at more than one point, then the graph does not represent a function.

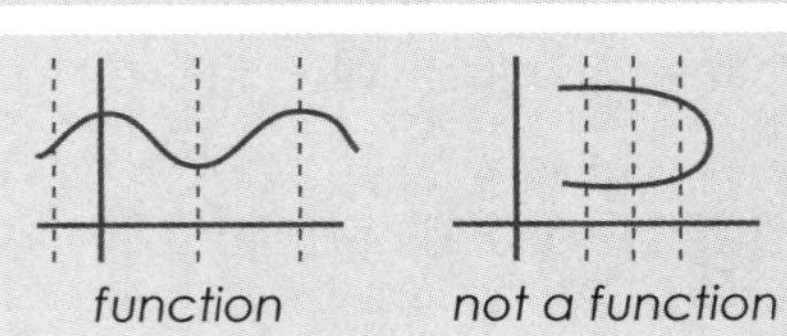

Vinculum (–)
The horizontal dash or "fraction line" separating the numerator and denominator of a fraction and which indicates division. [The forward slash "/" is also used for the same purpose, as in $^3/_4$.]

The "–" separating the 3 and the 4 in the fraction,
so that $\frac{3}{4} = 3 \div 4 = 0.75$

Volume (of a solid)
The amount of space that the solid occupies; capacity. Volume may be measured in cubic units.

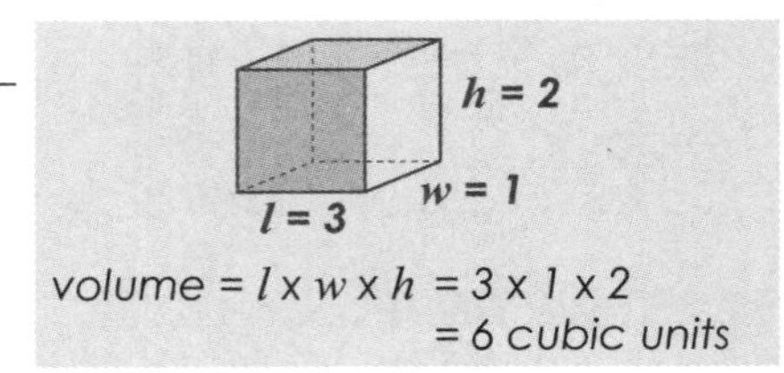

volume = $l \times w \times h = 3 \times 1 \times 2$
$= 6$ cubic units

Week
A unit of time consisting of seven days [see p. 121].

Weight
The force of gravity acting on an object; can be used to measure mass. [Weight is measured in pounds, ounces, and tons in the English system and in Newtons in the metric system, after the great mathematician Sir Isaac Newton, who developed the laws of gravity.]

Whole numbers (W)
The set W = {0, 1, 2, 3, 4, 5, ... }.

Width
See Length or Dimension.

x-axis

The horizontal axis in the coordinate plane.

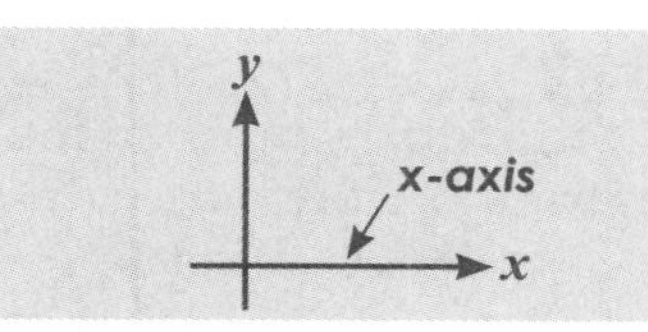

x-coordinate (Abscissa)

The first number in an ordered pair (x, y) representing a point in the coordinate plane.

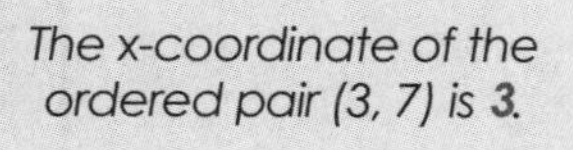

x-intercept

The x-coordinate of the point where the graph intersects the x-axis.

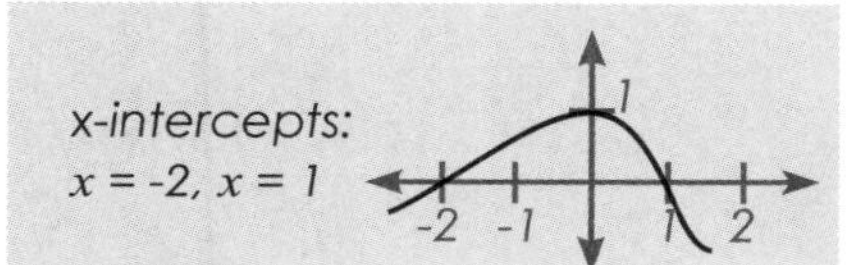

y-axis

The vertical axis in the coordinate plane.

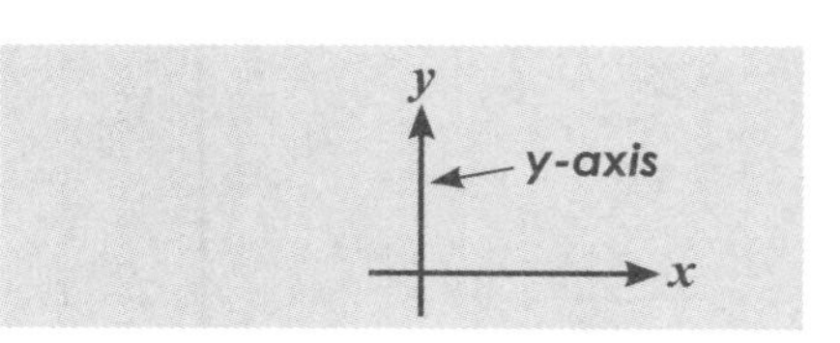

y-coordinate (Ordinate)

The second number in an ordered pair (x, y) representing a point in the coordinate plane.

The y-coordinate of the ordered pair (3, 7) is **7.**

y-intercept

The *y*-coordinate of the point where the graph intersects the *y*-axis.

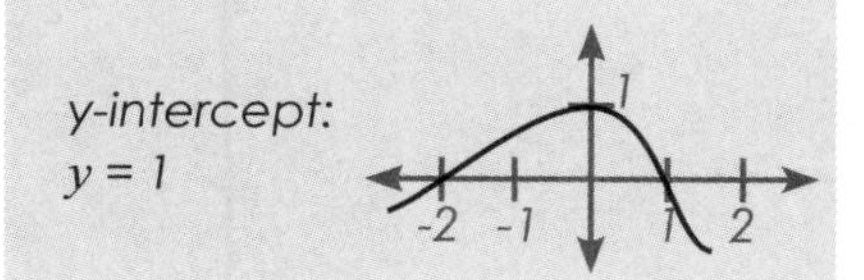

Yard

An English unit of length.

1 yard ≈ 91 cm

Year

The time it takes the Earth to revolve around the sun and set at 365 days for a normal year and 366 days in a leap year. [The exact time is 365 days, 5 hours, 48 minutes, 45 seconds; see p. 110.]

Z

Zero (0)

The cardinal number of the empty set. The first whole number. Zero is neither positive nor negative.

Additional Information

Tables

Number Systems

Charts

2-D Shapes

3-D Shapes

Measurement Systems

Conversion Tables

Equivalences

Formulas

Rules

Explanations

Symbols

Addition Table

+	0	1	2	3	4	5	6	7	8	9
0	0	1	2	3	4	5	6	7	8	9
1	1	2	3	4	5	6	7	8	9	10
2	2	3	4	5	6	7	8	9	10	11
3	3	4	5	6	7	8	9	10	11	12
4	4	5	6	7	8	9	10	11	12	13
5	5	6	7	8	9	10	11	12	13	14
6	6	7	8	9	10	11	12	13	14	15
7	7	8	9	10	11	12	13	14	15	16
8	8	9	10	11	12	13	14	15	16	17
9	9	10	11	12	13	14	15	16	17	18

Using the Table

For example, to find the sum 8 + 5, go along Row 8 and come down Column 5 to meet at 13 in the shaded area. To find the sum 5 + 8, go along Row 5 and come down Column 8 to get to 13 in the unshaded area.

Notice that the unshaded area has the same set of sums as on the shaded side of the diagonal, so the unshaded part of the table is not needed, because changing the order of addition still gives the same result: 8 + 5 = 5 + 8, by the Commutative Property of Addition.

Because addition and subtraction are opposite operations, the table also shows all the subtraction facts.

Except for the diagonal, every number in the table shows a family of four facts in one; e.g. 13 shows 8 + 5 = 13, 5 + 8 = 13, 13 – 5 = 8, and 13 – 8 = 5. Numbers in the diagonal show only two facts; e.g. 7 + 7 = 14, and 14 – 7 = 7.

X	0	1	2	3	4	5	6	7	8	9
0	0	0	0	0	0	0	0	0	0	0
1	0	1	2	3	4	5	6	7	8	9
2	0	2	4	6	8	10	12	14	16	18
3	0	3	6	9	12	15	18	21	24	27
4	0	4	8	12	16	20	24	28	32	36
5	0	5	10	15	20	25	30	35	40	45
6	0	6	12	18	24	30	36	42	48	54
7	0	7	14	21	28	35	42	49	56	63
8	0	8	16	24	32	40	48	56	64	72
9	0	9	18	27	36	45	54	63	72	81

Using the Table

For example, to find the product 8 x 5, go along Row 8 and come down Column 5 to meet at 40 in the shaded area. To find the product 5 x 8, go along Row 5 and come down Column 8 to get to 40 in the unshaded area.

Notice that the unshaded area has the same set of products as on the shaded side of the diagonal, so the unshaded part of the table is not needed, because changing the order of multiplication still gives the same result: 8 x 5 = 5 x 8 by the Commutative Property of Multiplication.

Because multiplication and division are inverse operations, the table also shows all the division facts.

Except for the diagonal, every nonzero number in the table shows a family of four facts in one; e.g. 40 shows 8 x 5 = 40, 5 x 8 = 40, 40 ÷ 5 = 8, and 40 ÷ 8 = 5. Numbers in the diagonal show only two facts; e.g. 7 x 7 = 49, and 49 ÷ 7 = 7.

Basic Properties of the Operations

Commutative Property of Addition

When adding two numbers, the order does not affect the sum: $a + b = b + a$ [see p. 70].

$7 + 3 = 3 + 7$

Commutative Property of Multiplication

When multiplying two numbers, the order does not affect the product: $a \times b = b \times a$ [See p. 71].

$7 \times 3 = 3 \times 7$

Associative Property of Addition

When adding three or more numbers, the grouping does not affect the sum: $(a + b) + c = a + (b + c)$. [So columns of numbers may be added up or down, or in convenient groups.]

$2 + (3 + 4) = (2 + 3) + 4$

Associative Property of Multiplication

When multiplying three or more numbers, the grouping does not affect the product: $(a \times b) \times c = a \times (b \times c)$. [So numbers can be multiplied in convenient groups.]

$2 \times (3 \times 4) = (2 \times 3) \times 4$

Distributive Property (of Multiplication Over Addition)

A number and a sum can be multiplied by multiplying by each part of the sum and then adding these products: $a \times (b + c) = (a \times b) + (a \times c)$. [The same property applies to multiplication over subtraction: $a \times (b - c) = (a \times b) - (a \times c)$.]

$$2 \times (3 + 7) = (2 \times 3) + (2 \times 7)$$

$$5 \times 967 = 5 \times (900 + 60 + 7) = (5 \times 900) + (5 \times 60) + (5 \times 7)$$

Addition Property of Zero (Identity Property of Addition)

When zero is added to any number, the sum is that number: $a + 0 = a$. [Zero is called the *additive identity*.]

$7 + 0 = 7$

Multiplication Property of Zero

The product of any number and zero is zero: $a \times 0 = 0$.

$7 \times 0 = 0$

Multiplication Property of One (Identity Property of Multiplication)

The product of any number and one is that number: $a \times 1 = a$. [One is called the *multiplicative identity*.]

$$7 \times 1 = 7$$

$$\frac{1}{2} = \frac{1}{2} \times 1 = \frac{1}{2} \times \frac{5}{5} = \frac{5}{10}$$

Decimal System of Numeration

The decimal system of numeration uses the ten digits 0, 1, 2, 3, 4, 5, 6, 7, 8 and 9 in a place value system. Each place in the system has a value ten times the value of the place on its right and one-tenth the value of the place on its left. All place values are powers of ten. Any real number, no matter how large or how small, can be represented in the system. The places are grouped in threes as shown below to assist in reading the number names.

Places		*Place values*
Hundred quadrillions	**18**	10^{17}
Ten quadrillions	**17**	10^{16}
Quadrillions	**16**	10^{15}
Hundred trillions	**15**	10^{14}
Ten trillions	**14**	10^{13}
Trillions	**13**	10^{12}
Hundred billions	**12**	10^{11}
Ten billions	**11**	10^{10}
Billions	**10**	10^{9}
Hundred millions	**9**	10^{8}
Ten millions	**8**	10^{7}
Millions	**7**	10^{6}
Hundred thousands	**6**	10^{5}
Ten thousands	**5**	10^{4}
Thousands	**4**	10^{3}
Hundreds	**3**	10^{2}
Tens	**2**	10^{1}
Units	**1**	10^{0}
	•	
Tenths	**1**	10^{-1}
Hundredths	**2**	10^{-2}
Thousandths	**3**	10^{-3}
Ten thousandths	**4**	10^{-4}
Hundred thousandths	**5**	10^{-5}
Millionths	**6**	10^{-6}

By using the triple groupings, the numeral 12,406,791 covers the place values of three sections—millions, thousands and units—and is read in three parts as "twelve million, four hundred six thousand, seven hundred ninety-one."

Place values to the right of the units or ones place are to represent numbers that are not integers. The number represented in the diagram below is 2,146.358, with the dot being a marker called a "decimal point" to separate the ones from the tenths. In reality, 2,146.358 means "two thousand one hundred forty-six and three hundred fifty-eight one-thousandths," but is read more briefly as 'two thousand, one hundred forty-six, point three five eight."

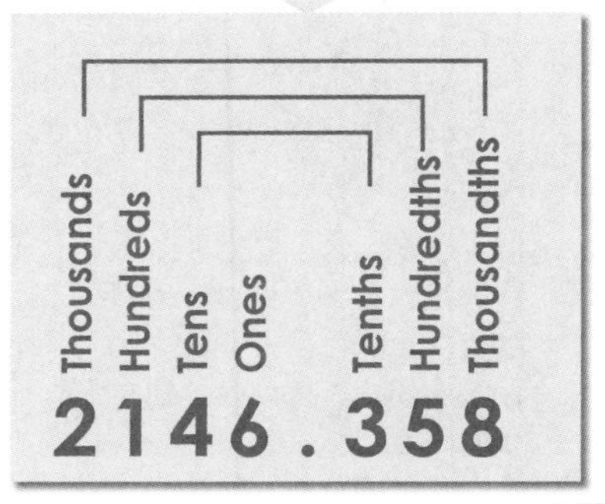

Roman Numeration System and Mayan Numeration System

Roman Letters Used In Numeration

Symbols used both in the Roman system of numeration and in our Hindu-Arabic system of numeration are shown below.

Roman	I	V	X	L	C	D	M
Hindu-Arabic	1	5	10	50	100	500	1000

Rules for using the Roman system

1. All symbols for decimal place values may be repeated up to three times.

III = 1 + 1 + 1 = 3;
XXX = 10 + 10 + 10 = 30;
CCC = 100 + 100 + 100 = 300;
MMM = 1000 + 1000 + 1000 = 3000.

2. A lesser value symbol in front of a greater value symbol means the lesser value is subtracted.

IV = 5 – 1 = 4;
IX = 10 – 1 = 9;
XL = 50 – 10 = 40;
CD = 500 – 100 = 400.

3. Lesser value symbols following a greater value symbol means the lesser value(s) is/are added.

VI = 5 + 1 = 6;
XV = 10 + 5 = 15;
LXIII = 50 + 10 + 3 = 63;
CXXXI = 100 + 10 + 10 + 10 + 1 = 131.

Mayan Symbols Used in Numeration

The Mayan numeration system was a modified base 20 positional system, with one of the first attempts at a symbol for zero.

Mayan	•	••	•••	••••	—	• —	— —	• — —
Hindu-Arabic	**1**	**2**	**3**	**4**	**5**	**6**	**10**	**11**

Mayan	•••• — — —	•	• •	• ••	••
Hindu-Arabic	**19**	**20**	**21**	**22**	**40**

In a base 20 numeration system the positions would be 1, 20, 20^2, 20^3, and so on. However, in the Mayan system, the positions were 1, 20, 18 x 20, 18 x 20 x 20, and so on.

Figurate Numbers

Figurate numbers are numbers that can be represented by arrangements of dots into the shapes of regular polygons. By following the patterns, each set of figurate numbers can be extended as far as required.

Triangular Numbers

Triangular numbers are those that can be represented by dots arranged in the shape of an equilateral triangle. The general formula for a triangular number is $\frac{n(n+1)}{2}$, where n is a counting number.

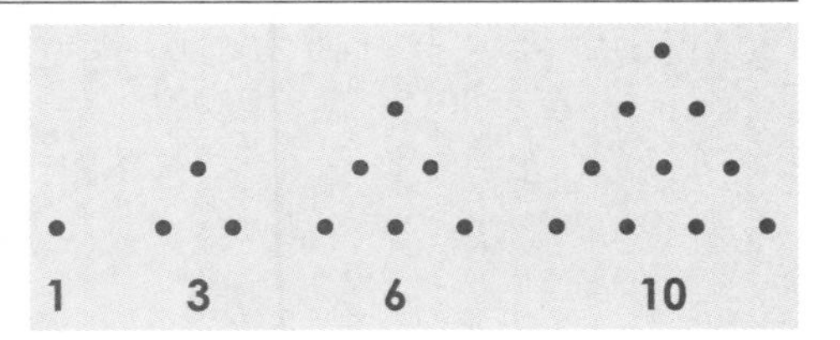

These can also be shown as:

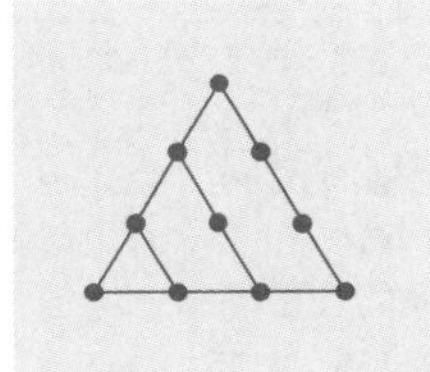

Square Numbers

Square numbers are those that can be represented by dots arranged in the shape of a square. These numbers can be obtained by multiplying a number by itself. The general formula for a square number is n^2, where n is a counting number.

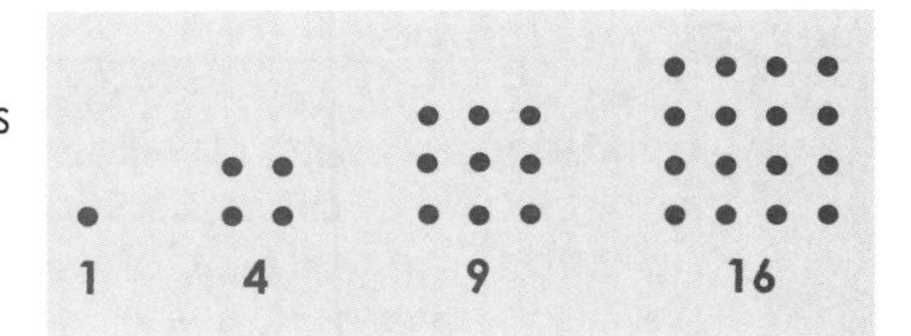

These can also be shown as:

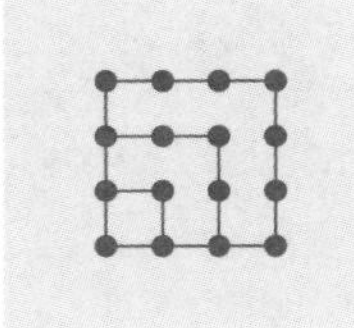

Relationship Between Triangular Numbers and Square Numbers

When any two consecutive triangular numbers are joined, they form a square number.

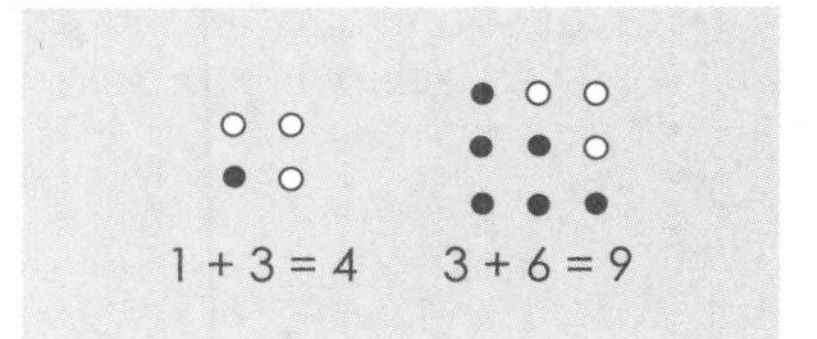

Figurate Numbers

Pentagonal Numbers

Pentagonal numbers are those that can be represented by dots arranged in the shape of a pentagon or regular pentagon. The general formula for a pentagonal number is $\frac{n(3n-1)}{2}$ where n is a counting number.

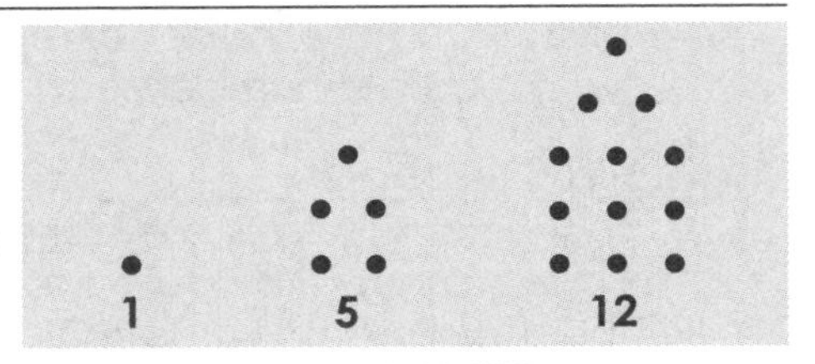

These can also be shown as:

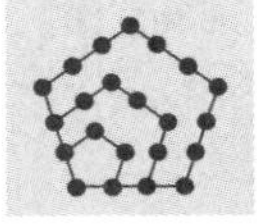

Relationship Between Pentagonal Numbers and Triangular And Square Numbers

When certain square numbers and triangular numbers are combined, they form pentagonal numbers. For example, the third pentagonal number is the sum of the third square number and the second triangular number.

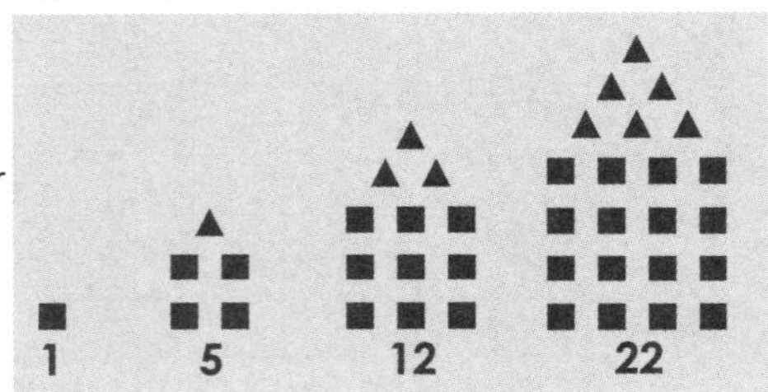

Hexagonal Numbers

Hexagonal numbers are those that can be represented by dots arranged in the shape of a hexagon or regular hexagon. The general formula for a hexagonal number is $n(2n - 1)$, where n is a counting number.

These can also be shown as:

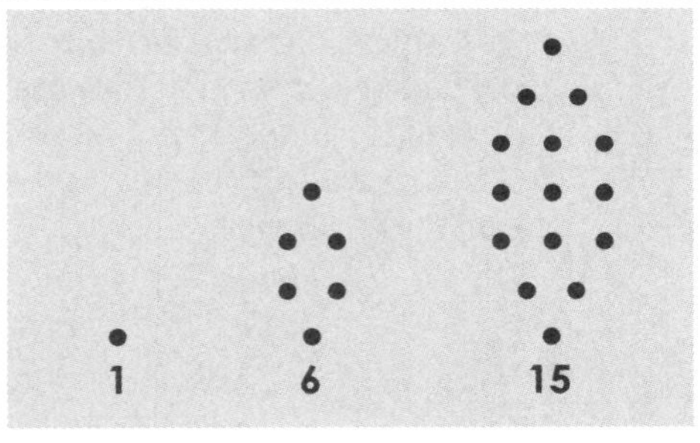

Relationship Between Hexagonal Numbers and Triangular and Square Numbers

When certain square numbers and triangular numbers are combined, they form hexagonal numbers. For example, the third hexagonal number is the sum of the third square number and twice the second triangular number.

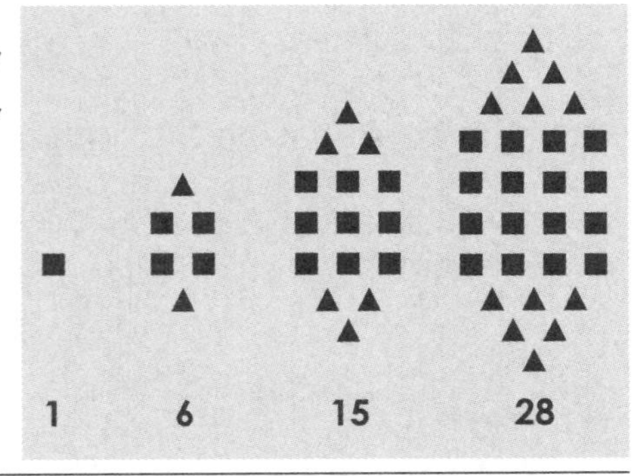

Figurate Numbers

TYPE	1st	2nd	3rd	4th	5th	6th	7th	8th	9th	10th
Triangular	1	3	6	10	15	21	28	36	45	55
Square	1	4	9	16	25	36	49	64	81	100
Pentagonal	1	5	12	22	35	51	70	92	117	145
Hexagonal	1	6	15	28	45	66	91	120	153	190
Heptagonal	1	7	18	34	55	81	112	148	189	235
Octagonal	1	8	21	40	65	96	133	176	225	280
Nonagonal	1	9	24	46	75	111	154	204	261	325
Decagonal	1	10	27	52	85	126	175	232	297	370

Factors

Any counting number that divides another without a remainder is a **factor** of that number. For example, 1, 2, 3, 5, 6, 10, 15 and 30 are factors of 30 because they all divide 30 without any remainder.

A prime number is a counting number greater than 1 that has two distinct factors, 1 and itself. A prime number that divides a given counting number without any remainder is called a **prime factor**. In the above example, 2, 3 and 5 are prime factors of 30.

Methods for Finding Prime Factors

1. Use a factor tree.

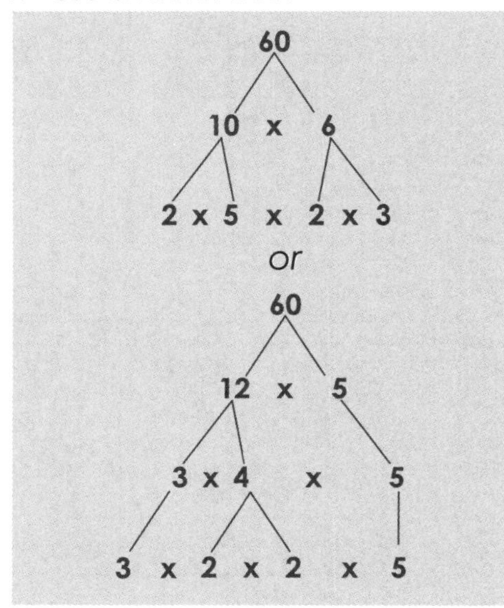

2. Divide by prime numbers and continue as much as possible.

$2\overline{)60}$ = 30

$2\overline{)30}$ = 15

$3\overline{)15}$ = 5

By either method, the prime factors of 60 are 2, 3, and 5 and the prime factorization of 60 is $2^2 \times 3 \times 5$.

Factors and Rectangular Numbers

All counting numbers can be represented by arrangements of dots in arrays, and the arrays show the factors of the numbers. Prime numbers have only two arrays and only two factors, whereas composite numbers have more than two arrays and more than two factors. Note that 1, with only one array is neither prime nor composite. The composite numbers are rectangular numbers since they can be represented by a rectangular array of dots with more than one row and more than one column. The arrays for the first 12 counting numbers are shown below.

	1	•			
Prime	**2**	••	• •		
Prime	**3**	•••	• • •		
Composite	**4**	••••	• • • •	•• ••	
Prime	**5**	•••••	• • • • •		
Composite	**6**	••••••	• • • • • •	•• •• ••	••• •••
Prime	**7**	•••••••	• • • • • • •		
Composite	**8**	••••••••	• • • • • • • •	•• •• •• ••	•••• ••••

Composite	9						
Composite	10						
Prime	11						
Composite	12						

Prime and Composite Numbers to 200

A counting number greater than 1 is either prime or composite. The prime numbers are shown in bold. The others, except 1, are composite numbers.

1	**2**	**3**	4	**5**	6	**7**	8	9	10
11	12	**13**	14	15	16	**17**	18	**19**	20
21	22	**23**	24	25	26	27	28	**29**	30
31	32	33	34	35	36	**37**	38	39	40
41	42	**43**	44	45	46	**47**	48	49	50
51	52	**53**	54	55	56	57	58	**59**	60
61	62	63	64	65	66	**67**	68	69	70
71	72	**73**	74	75	76	77	78	**79**	80
81	82	**83**	84	85	86	87	88	**89**	90
91	92	93	94	95	96	**97**	98	99	100
101	102	**103**	104	105	106	**107**	108	**109**	110
111	112	**113**	114	115	116	117	118	119	120
121	122	123	124	125	126	**127**	128	129	130
131	132	133	134	135	136	**137**	138	**139**	140
141	142	143	144	145	146	147	148	**149**	150
151	152	153	154	155	156	**157**	158	159	160
161	162	**163**	164	165	166	**167**	168	169	170
171	172	**173**	174	175	176	177	178	**179**	180
181	182	183	184	185	186	187	188	189	190
191	192	**193**	194	195	196	**197**	198	**199**	200

Rules for Divisibility

Divisibility	Rule
Two	The last digit is 0, 2, 4, 6, or 8, For example, 70, 240, 3,964 and 11,486.
Three	The sum of the digits of the number is divisible by 3. For example, 192 is divisible by 3, since 1 + 9 + 2 = 12 is divisible by 3.
Four	The number represented by the last two digits is divisible by 4. For example, 528, 3,228 and 13,028 are all divisible by 4, since 28 is divisible by 4.
Five	The last digit of the number is either 0 or 5. For example, 65, 740, 935 and 3,890 all end in either 0 or 5 so they are all divisible by 5.
Six	The number is even and the sum of its digits is divisible by 3. For example, 732 is even and 7 + 3 + 2 = 12 which is divisible by 3, so 732 is divisible by 6.
Seven	The double of the last digit subtracted from the remaining number is divisible by 7. For example, 1,666: 2 x 6 = 12 → 166 – 2 = 154 → 2 x 4 = 8 → 15 – 8 = 7, and since 7 is divisible by 7, so is 1,666. Caution: The rule is not efficient.
Eight	The number represented by the last three digits is divisible by 8. For example, 5,936, 24,936 and 125,936 are all divisible by 8, since 936 is divisible by 8.
Nine	The sum of the digits is divisible by 9. For example, 738 is divisible by 9, since 7 + 3 + 8 = 18 is divisible by 9.
Ten	The last digit of the number is 0. For example, 90, 170 and 3,710 are all divisible by 10, since they all end in 0.

Rational Number Equivalents

Fraction	Decimal	Percent
$\frac{1}{1}$	1	100%
$\frac{1}{2}$	0.5	50%
$\frac{1}{3}$	$0.\overline{3}$	$33.\overline{3}\%$
$\frac{2}{3}$	$0.\overline{6}$	$66.\overline{6}\%$
$\frac{1}{4}$	0.25	25%
$\frac{3}{4}$	0.75	75%
$\frac{1}{5}$	0.2	20%
$\frac{2}{5}$	0.4	40%
$\frac{3}{5}$	0.6	60%
$\frac{4}{5}$	0.8	80%
$\frac{1}{6}$	$0.1\overline{6}$	$16.\overline{6}\%$
$\frac{5}{6}$	$0.8\overline{3}$	$83.\overline{3}\%$
$\frac{1}{8}$	0.125	12.5%
$\frac{3}{8}$	0.375	37.5%
$\frac{5}{8}$	0.625	62.5%
$\frac{7}{8}$	0.875	87.5%
$\frac{1}{10}$	0.1	10%
$\frac{3}{10}$	0.3	30%
$\frac{7}{10}$	0.7	70%
$\frac{9}{10}$	0.9	90%
$\frac{1}{12}$	$0.08\overline{3}$	$8.\overline{3}\%$
$\frac{1}{20}$	0.05	5%
$\frac{1}{25}$	0.04	4%
$\frac{1}{40}$	0.025	2.5%
$\frac{1}{50}$	0.02	2%
$\frac{1}{100}$	0.01	1%
$\frac{1}{1{,}000}$	0.001	0.1%
$\frac{1}{10{,}000}$	0.0001	0.01%
$\frac{67}{100}$	0.67	67%

The Real Number System

The chart below classifies sets of numbers. The first set used is the set of *counting numbers* or *natural numbers* N = {1, 2, 3, 4 ...}. The counting numbers, with *zero*, make up the *whole numbers* W = {0, 1, 2, 3, 4 ...}. The whole numbers together with their opposites, $^-1$, $^-2$, $^-3$, ... make up the set of *integers* I = {... $^-3$, $^-2$, $^-1$, 0, 1, 2, 3, ...}. Then there are the *non-integer rational numbers* such as $\frac{-3}{4}$, $\frac{2}{3}$, $\frac{7}{5}$, and $\frac{-5}{2}$; that together with the integers form the set of *rational numbers* Q. The *irrational numbers* Q', such as $\sqrt{2}$, $^-\sqrt{5}$, π, and $^-\pi$, together with the rational numbers form the set of *real numbers* where R = Q ∪ Q'.

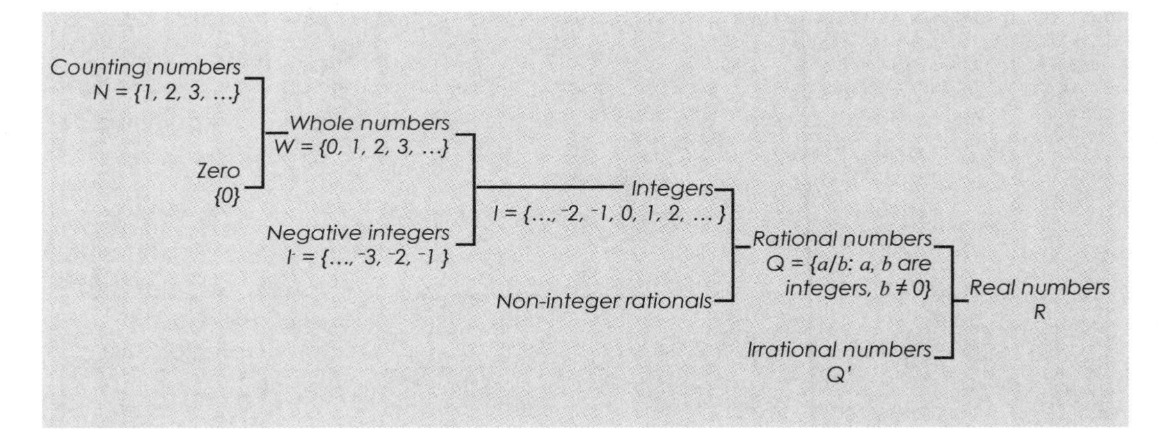

Note: Rational numbers are based on ratios and are all numbers resulting from the division of any integer by any other integer except 0. Thus all rational numbers can be represented by a fraction, while irrational numbers cannot be expressed in this way but are still represented on the number line. Note how $\sqrt{2}$ is located there. Thus the set of real numbers fills all points on the number line.

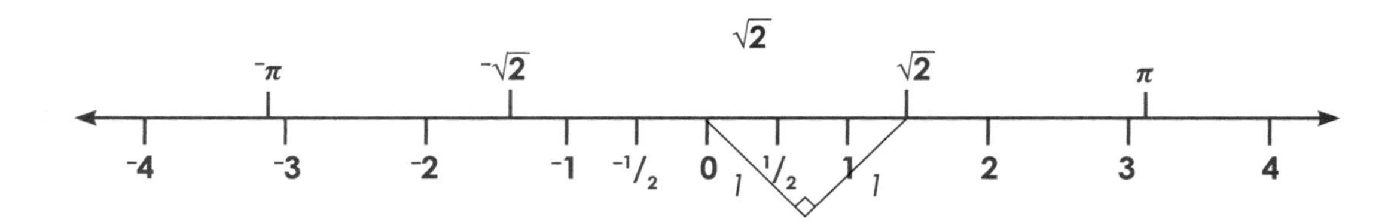

Exponents

A quick way of writing 3 x 3 x 3 x 3 is 3^4. In this case 3 is called the **base**, and 4 is called the **exponent** (power). The exponent indicates the power to which the base has been raised. The number 3^4 is read as "three raised to the power four," or as "three to the fourth power," or more briefly as "three to the fourth." Note: The term "power" can describe the whole expression 3^4 or the exponent 4, depending on context. The exponent rules for multiplication and division are detailed below.

Exponent Rule	Example	General Rule
Multiplication		
When multiplying powers with the same base, add the exponents.	$3^2 = 3 \times 3$ $3^4 = 3 \times 3 \times 3 \times 3$ $3^2 \times 3^4 = 3 \times 3 \times 3 \times 3 \times 3 \times 3$ $= 3^6$ $= 3^{2+4}$	If a is any real number and m and n are positive integers, then $a^m \times a^n = a^{m+n}$
Division		
When dividing powers with the same base, the exponent of the denominator is subtracted from the exponent of the numerator.	$\frac{3^6}{3^2} = \frac{3 \times 3 \times 3 \times 3 \times 3 \times 3}{3 \times 3}$ $= 3^{6-2}$ $= 3^4$	If a is any real number and m and n are positive integers, then $\frac{a^m}{a^n} = a^{m-n}$
Power of zero		
Any real number, except 0, raised to the power zero has a value of 1.	$\frac{3^4}{3^4} = 3^{4-4}$ $= 3^0$ $= 1$	If a is any nonzero real number, then $a^0 = 1$
Power raised to powers		
When a base is raised to a power, and the result is raised to another power, multiply the exponents.	$(3^2)^3 = (3 \times 3)^3$ $= 3 \times 3 \times 3 \times 3 \times 3 \times 3$ $= 3^{2 \times 3}$ $= 3^6$	If a is any real number and m and n are positive integers, then $(a^m)^n = a^{mn}$

Rules of Exponents

Exponent Rule	Example	General Rule
Powers of products		
When a product is raised to a power, every factor of the product is raised to that power.	$(3 \times 5)^2 = (3 \times 5) \times (3 \times 5)$ $= 3^2 \times 5^2$	If a and b are any real numbers and n is a positive integer, then $(ab)^n = a^n b^n$
Powers of quotients		
When a quotient is raised to a power, both the numerator and the denominator are raised to that power.	$(\frac{3}{5})^3 = \frac{3}{5} \times \frac{3}{5} \times \frac{3}{5}$ $= \frac{3^3}{5^3}$	If a and b are any real numbers, $b \neq 0$, and n is a positive integer, then $(\frac{a}{b})^n = \frac{a^n}{b^n}$
Negative powers		
Raising any real number except 0 to a negative power results in the reciprocal.	$3^{-2} = \frac{1}{3^2}$	If a is any nonzero real number and n is a positive integer, then $a^{-n} = \frac{1}{a^n}$
Fractional powers		
When an exponent is a fraction, the denominator indicates the root.	$5^{\frac{2}{3}} = \sqrt[3]{5^2}$	If a is any nonzero real number and m and n are positive integers, then $a^{\frac{m}{n}} = \sqrt[n]{a^m}$

Mathematics of Finance

Simple Interest

Simple interest is the amount of money paid on an invested amount called the principal (P), calculated at an annual percentage rate (r), and paid at the end of a period of time (t) expressed in years.

To find the simple interest on $800 invested at 7% (r = .07) for two years:

$$\begin{aligned} \textbf{Interest} &= P \times r \times t \\ &= \$800 \times .07 \times 2 \\ &= \$112 \end{aligned}$$

Compound Interest

When the interest earned on an investment is added to the principal each year, the resulting interest is called **compound interest**.

To find the compound interest on $800 invested at seven percent for two years:

$$\begin{aligned} \textbf{Year 1: Balance} &= \text{Principal} + \text{Interest} \\ &= \$800 \times (\$800 \times .07) \\ &= \$800 \times (1.07) \end{aligned}$$

$$\begin{aligned} \textbf{Year 2: Balance} &= \text{Principal} + \text{Interest} \\ &= \$800\ (1.07) + \$800\ (1.07)\ (.07) \\ &= \$800\ (1.07)^2 = \$915.92 \\ \textbf{Interest} &= \$915.92 - \$800 = \$115.92 \end{aligned}$$

If principal P is invested at a rate r, compounded annually for t years, then:

$$\begin{aligned} \textbf{Balance} &= P\ (1 + r)^t \\ \textbf{Interest} &= P\ (1 + r)^t - P \end{aligned}$$

Commission

A **commission** is a payment made for a service and calculated as a percentage of the total amount of the transaction.

If a car salesperson is paid 2% (.02) commission for selling a car for $12,000, then the commission is calculated as follows:

$$\begin{aligned} C &= .02 \times \$12{,}000 \\ &= \$240 \end{aligned}$$

Discount

A **discount** is the amount taken off the price of an item and is expressed either as an amount of money or as a percentage off the marked price.

If the marked price of an item is $48 and the discount is 10% (.10), the amount of money taken off is calculated as follows:

$$\text{Discount} = .10 \times \$48 = \$4.80$$

$$\text{Final selling price} = \$48 - \$4.80 = \$43.20$$

Profit and Loss

Profit is the amount of money gained in a transaction or over a certain period of time, such as a year; if this amount is negative it is called a **loss**. Profit can be expressed both as an amount of money or as a percentage of the transaction(s).

If a store buys an item at a wholesale cost of $250 and fixes the retail price at $300, then the profit and profit rate or profit margin from the sale is calculated as follows:

Retail Price = $300 **Wholesale = $250**

$$\text{Profit} = \text{Retail Price} - \text{Wholesale Cost} = \$300 - \$250 = \$50$$

$$\text{Profit Rate} = \frac{\text{Profit}}{\text{Wholesale Cost}} = \frac{\$50}{\$250} \times 100\% = .20 \text{ or } 20\%$$

Sales Tax

In the United States, most states apply a **sales tax** to purchases, though tax rates vary.

The State of Indiana applies a 6% sales tax to purchases. If the original cost of an item is $65, then the final cost (including tax) is calculated as follows:

$$\text{Final cost} = \text{Original cost} + \text{Sales Tax} = \$65 + .06\ (\$65) = \$65 + \$3.90 = \$68.90$$

Data Representation

Carroll Diagram

A grid-like structure for categorizing results. The example below shows colored blocks sorted into a 2 by 2 grid.

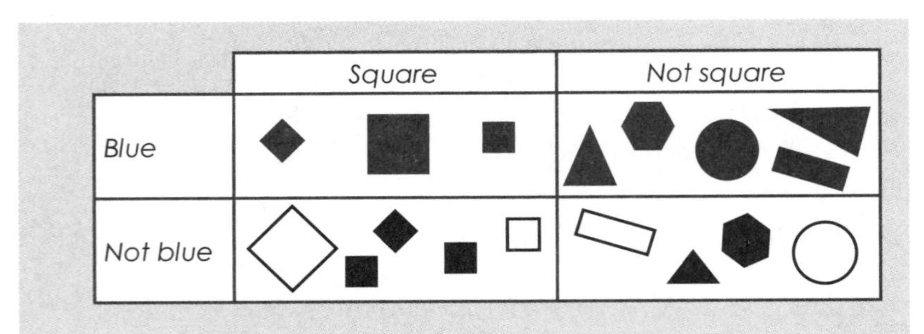

Tree Diagram

A branching diagram that shows all the possible choices or outcomes of a process carried out in several stages.

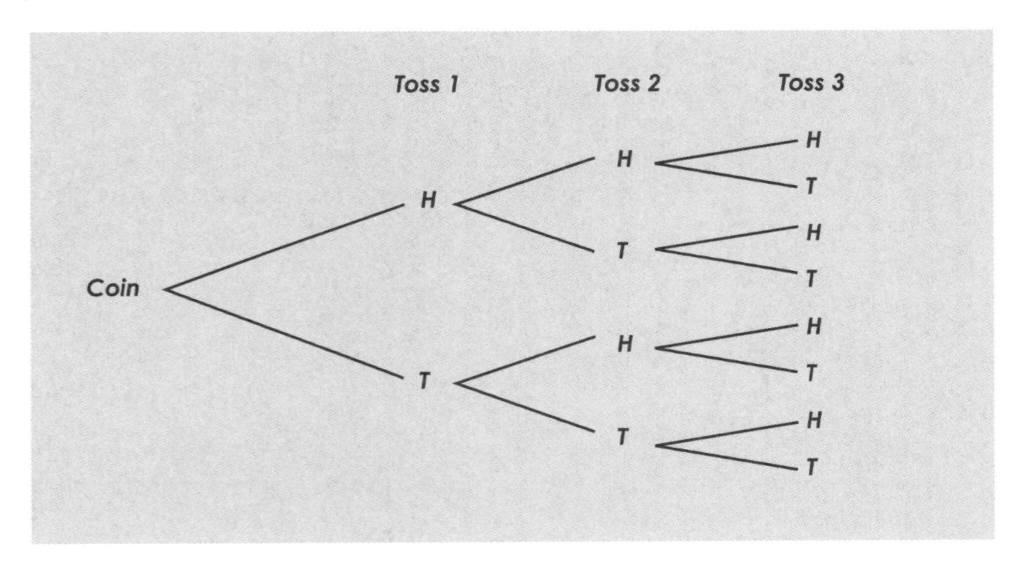

Venn Diagram

A diagram that represents sets and their relationships. The example below shows the same colored blocks as in the Carroll diagram sorted using a Venn diagram.

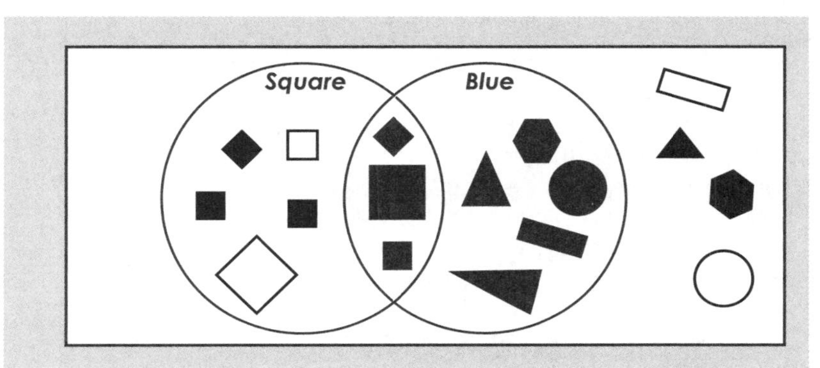

Table of Data

A table is a means of organizing data in rows and columns for a particular purpose. The example below shows the data for the heights of a class of students.

Height in centimeters of 11-year-old students

Height in cm	Tally	Frequency
108	|	*1*
112	|||	*3*
115	||	*2*
119	||||	*4*
120	|||	*3*
122	||||	*4*
123	~~||||~~ |	*6*
126	||	*2*
130	|	*1*
132	|||	*3*
137	|	*1*

Measures of Central Tendency

There are three common measures of central tendency for a set of data, such as heights of students.

Mode

From this table we can see that the modal or most commonly occurring height is 123 cm. [Note that sometimes there may be more than one mode; e.g. if there were only three students who were 123 cm tall, the heights of 119 cm and 122 cm would be the most commonly occurring ones, so the data would be bi-modal, or have two modes.]

Height in cm	Tally	Frequency
108	|	*1*
112	|||	*3*
115	||	*2*
119	||||	*4*
120	|||	*3*
122	||||	*4*
123	~~||||~~ |	*6*
126	||	*2*
130	|	*1*
132	|||	*3*
137	|	*1*

Median

The median (or middle) height may be found by locating the midpoint in the data when the data is arranged in order. Arranging the heights in order from shortest to tallest:

108, 112, 112, 112, 115, 115, 119, 119, 119, 119, 120, 120, 120, 122, 122, 122, 122, 123, 123, 123, 123, 123, 123, 126, 126, 130, 132, 132, 132, 137

So the median height is 122 cm, as there are 13 heights less and 13 heights more than 122.

If there were 29 students, the middle height or 15th height would be the median.

If there were only 28 students, the median would be between the 14th and 15th heights—thus the mean of the 14th and 15th heights.

Mean

The mean or average height is found by adding the heights of all the students and dividing by the number of students. In the above example, the total of all the heights added together is 3651, and this is divided by the number of students, 30; i.e. mean = 3651 cm ÷ 30 = 121.7 cm.

Bar Graph (Column Graph)

A visual way to represent a set of data by bars (rectangles) of equal width. The length of each bar/column corresponds to the size or frequency. Bars may be vertical or horizontal. Usually the bars or columns only touch each other if the data is continuous; and for discrete data, there is a space left between each bar or column.

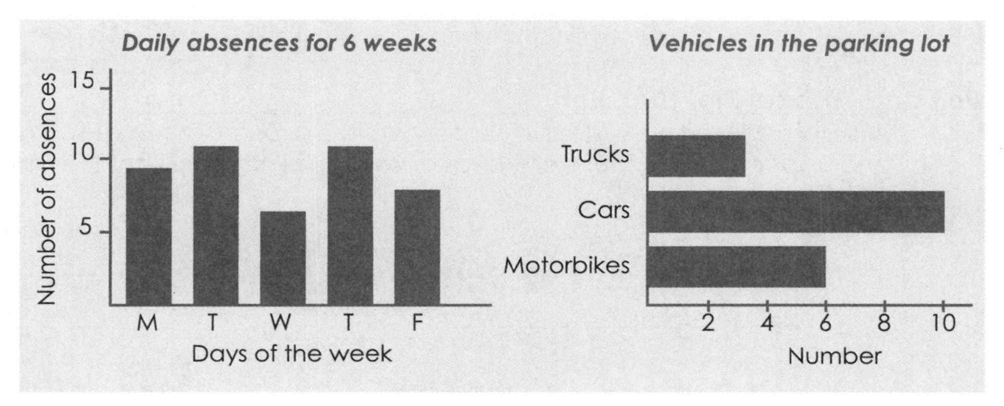

A bar graph may be used to represent more than one set of data. This is achieved by subdividing the bars to represent the components [known as a component, composite, stacked or sectional bar graph]; or the data can be displayed with bars alongside each other [known as a compound or multiple bar graph]. Examples are shown below.

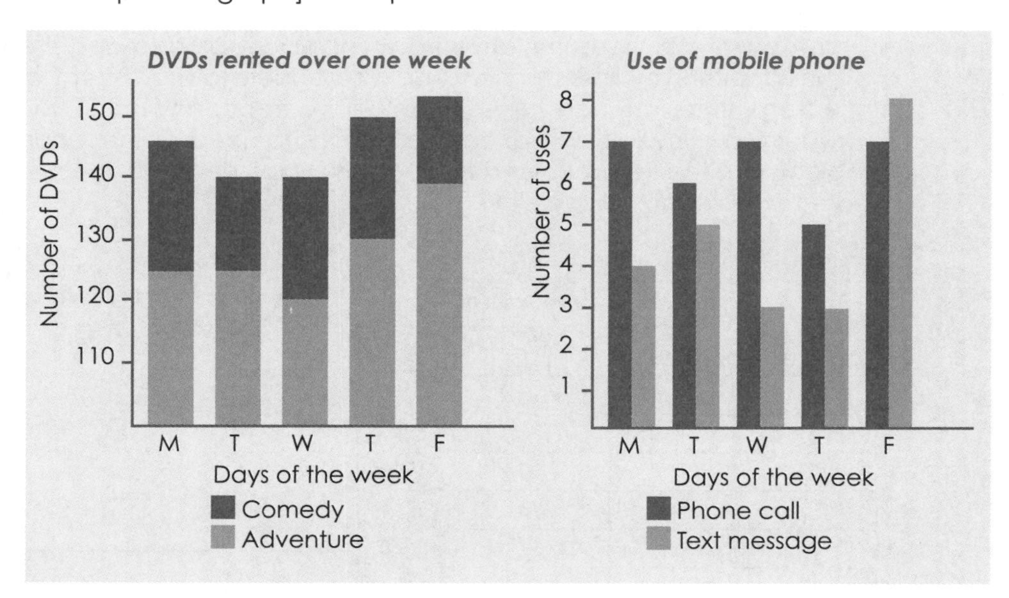

Data Representation

Block Graph

A type of bar or column graph where each bar or column is divided to show all individual pieces of data.

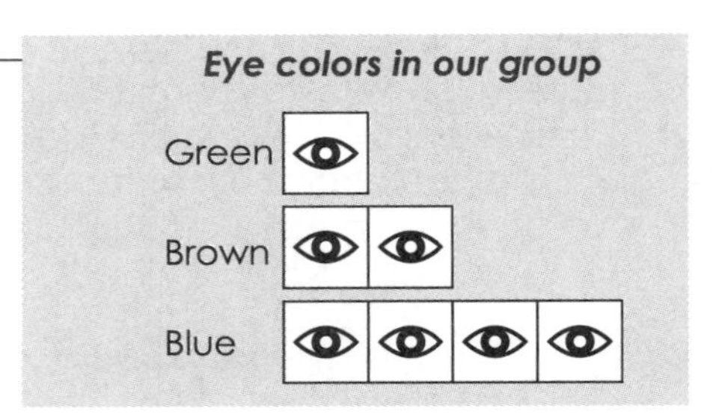

Box and Whisker Plot (Boxplot)

A graphical summary of data that shows five aspects of the data: the lower and upper quartiles (hence inter-quartile range), the median and the lowest and highest values.

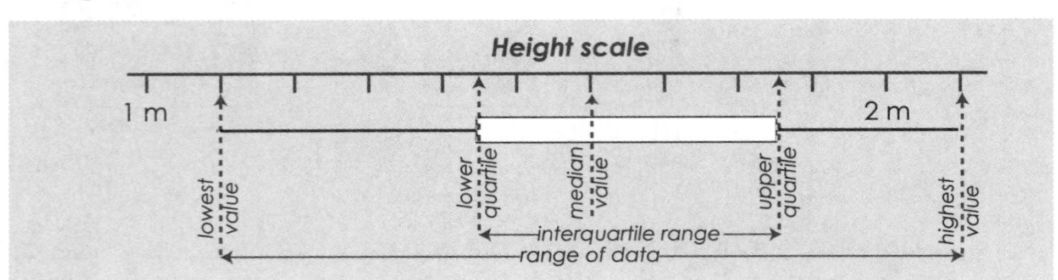

Column Graph

See Bar Graph.

Cumulative Frequency Graph (Ogive)

A graph on which the cumulative frequencies are plotted and the points are joined, either by line segments or, where there are sufficient points, a curve (sometimes referred to as a cumulative frequency curve, or ogive). To produce a cumulative frequency graph, data needs to be recorded on a table where the frequencies are successively totalled; e.g. the number of children couples have, as below.

Number of children	Number of couples (f)	Cumulative frequency (cf)
0	4	4
1	7	11
2	24	35
3	9	44
4	5	49
5	1	50

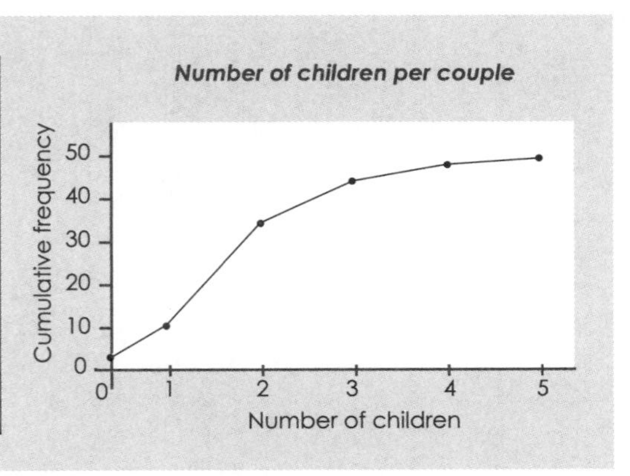

Dot Plot (Line Plot)

A data display consisting of a horizontal number line on which each data point is denoted by a dot or an X above the corresponding number line value.

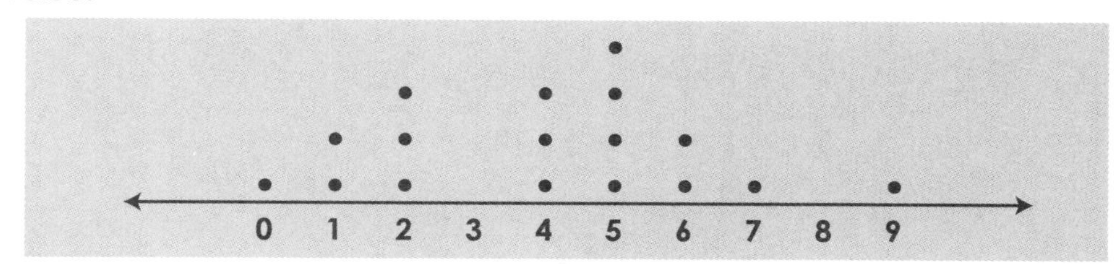

Histogram

A bar or column graph with the added factor that the width of each bar or column corresponds to an interval of values on the horizontal axis, and the height of each bar indicates the frequency of data. There are no gaps between the bars, and when the width of each bar is the same, it looks like a bar graph.

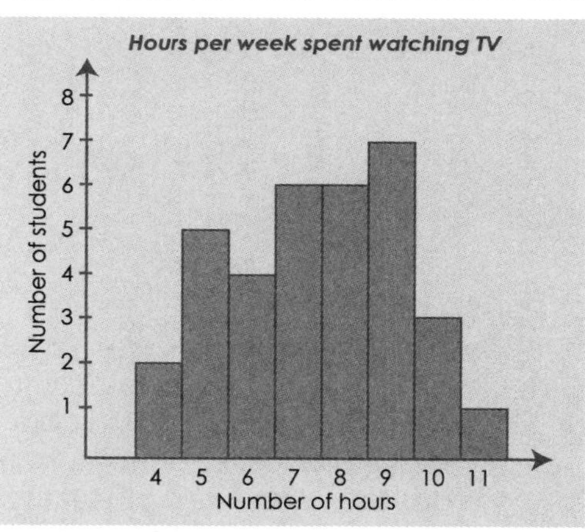

Some of the differences between a histogram and a bar graph

Histogram	Bar Graph
Only used to show how frequently some quantity occurs.	*May be used to show information other than frequency.*
Does not have gaps between the columns.	*Usually has gaps between the columns.*
Bars are vertical.	*Bars may be horizontal or vertical.*
The height of each column represents the frequency.	*The length of each rectangle represents the frequency.*
The arrangement of the columns is important as the horizontal axis represents values that have been counted or measured. The columns are placed so that the categories of data they represent are in numerical order.	*The order of the columns is not generally important.*
Categories of data with no actual data are shown on the scale.	*Categories of data with no actual data are not necessarily shown on the scale.*

Frequency Polygon

A graph formed by joining the midpoints of the tops of the bars of a histogram by line segments, which are connected at the start and finish of the graph to produce a polygon.

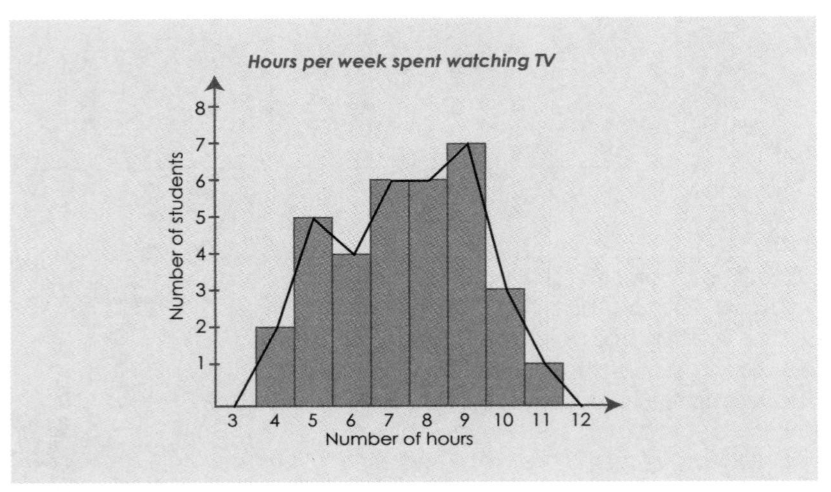

Line Graph

A graph formed by line segments connecting points representing certain data, and with the horizontal axis usually indicating a measure of time; normally used for continuous or measurement data. Values can only be read accurately from the marked points, although estimates can be made from other positions along the line segments (interpolation); or beyond the points (extrapolation), which is more speculative.

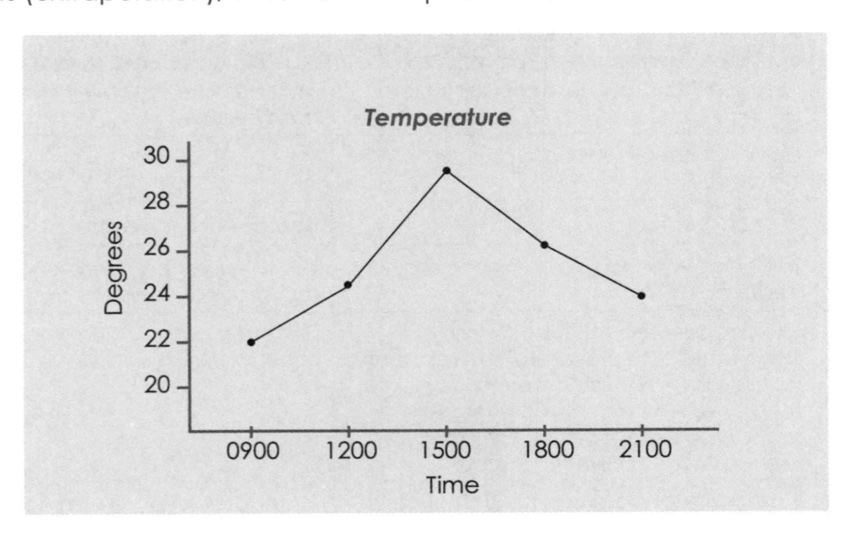

Pictograph (Picture Graph)

A graph that uses pictures or symbols to display data, where one picture may represent one or more units. When different pictures are used, they should be aligned on the chart.

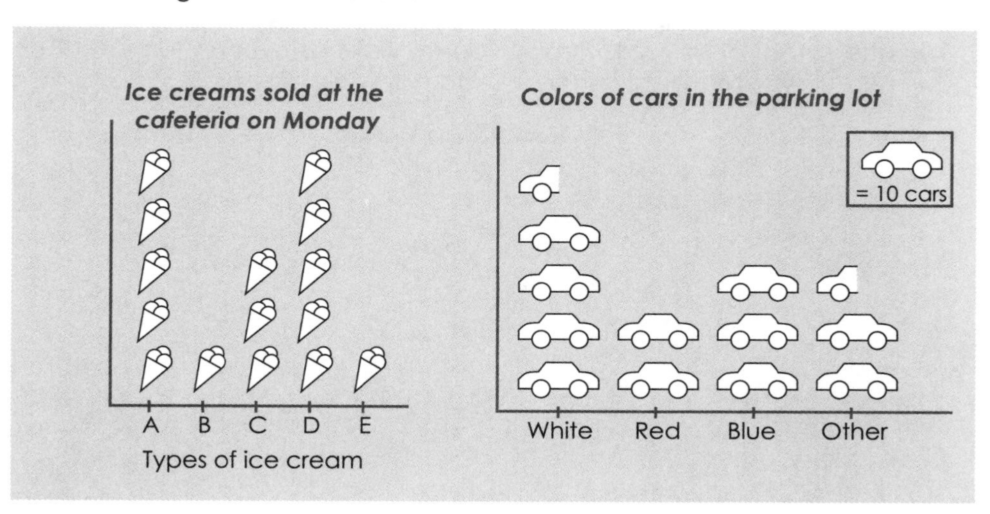

Pie Chart (Pie Graph; Circle Graph)

The sectors of a circle are used to show a whole in terms of its parts.

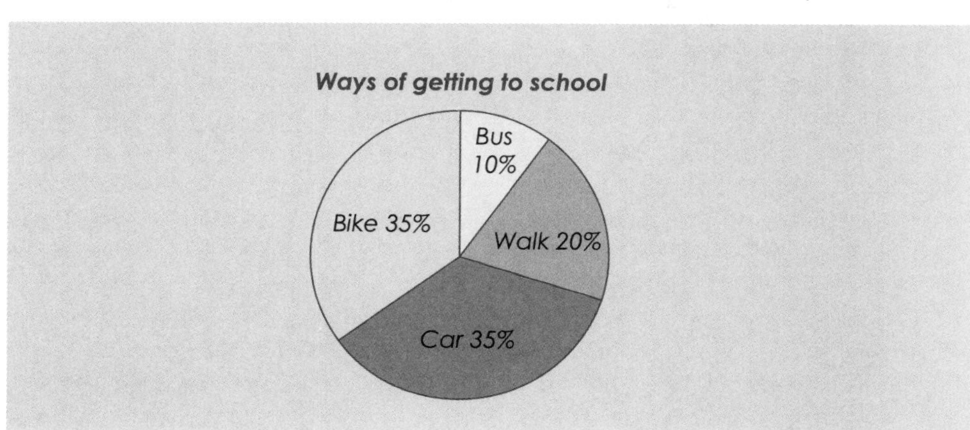

Scatterplot (Scattergram)

A graph of plotted points which display the relationship between pairs of data sets. It uses bivariate data, which is information collected for the purpose of comparison. ["Bivariate" meaning two variables—most graphs referred to previously have been univariate—one variable.] Used to indicate the extent of any relationship (correlation) between the pairs of data sets, which is determined by the spread pattern of the plots.

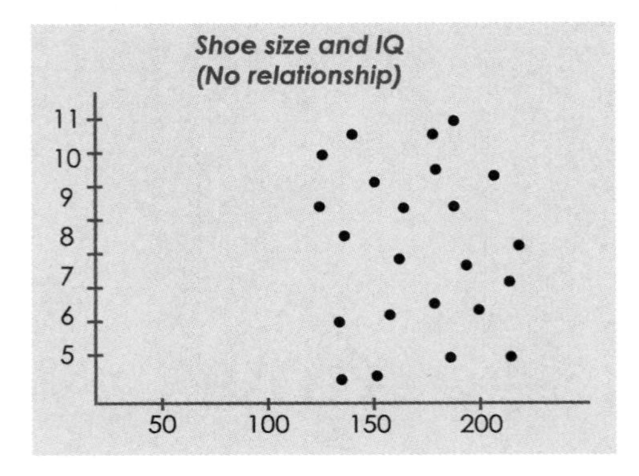

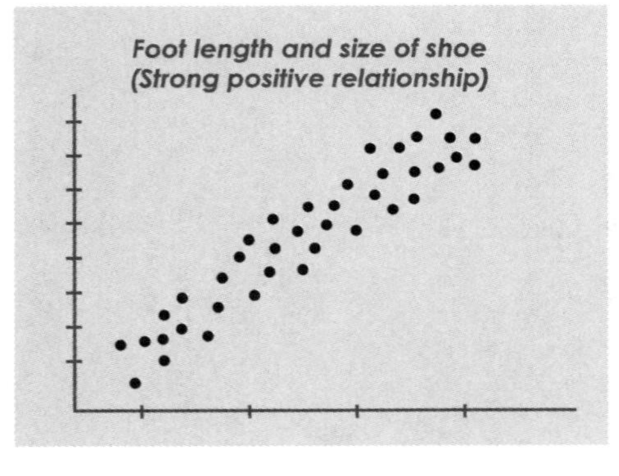

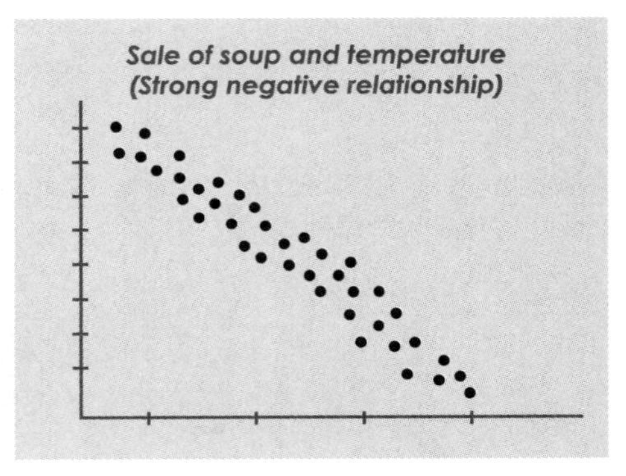

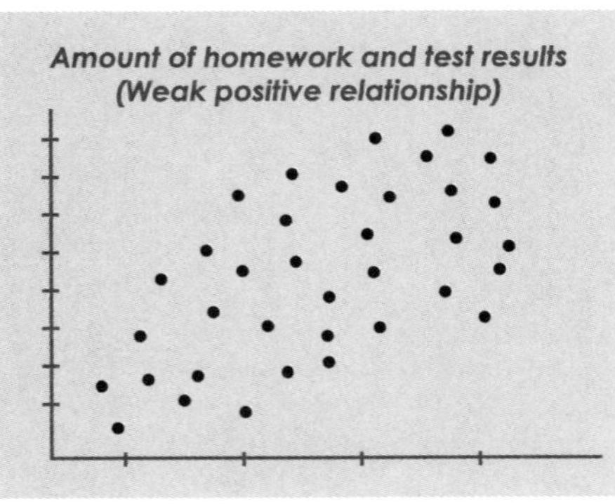

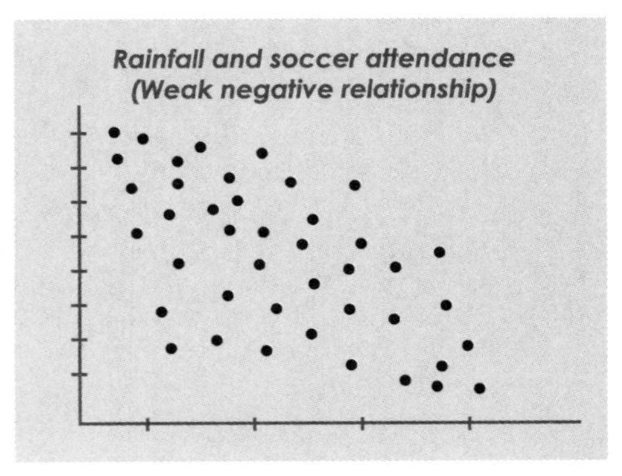

Stem-and-Leaf Plot

Display of data with its frequency, using part of the value of the data on one side of the plot (the stem); and the remainder of the value on the other side of the plot (the leaf).

Scores in a game

Stem	Leaves
1	9 6 9
2	1 7 6 4 8
3	4 5 2

So the scores were: 19, 16, 19, 21, 27, 26, 24, 28, 34, 35 and 32.

Comparison of male and female pulse rates

Females		Males
0	10	
2 8	9	2
4 6 5 8 1	8	6 3 5
5 0 7 6	7	3 5 4 2 3 4
3 5 8 1 3	6	0 7 8 5
5	5	2 9

Reading from the top the female pulse rates were 100, 92, 98, 84, 86 etc. The male pulse rates were 92, 86, 83, 85, 73, etc.

Angles

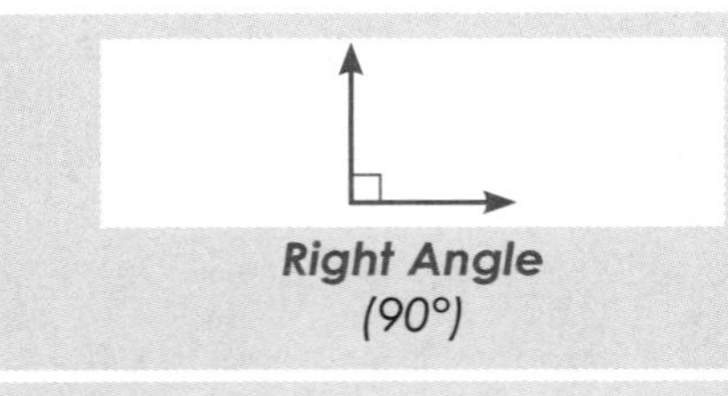

Right Angle
(90°)

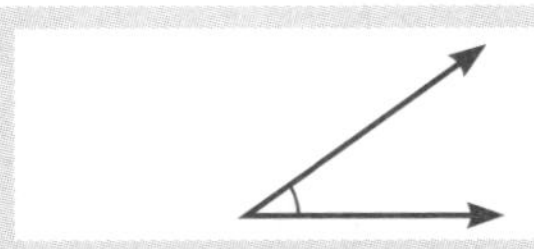

Acute Angle
(greater than 0° and less than 90°)

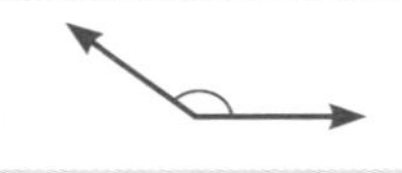

Obtuse Angle
(greater than 90° and less than 180°)

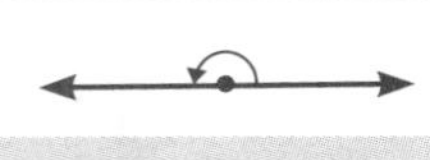

Straight Angle
(180°)

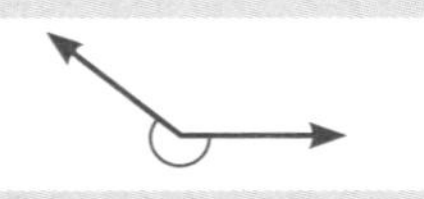

Reflex Angle
(greater than 180° and less than 360°)

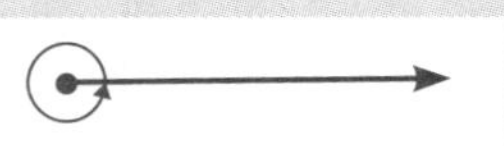

One Rotation (Full Turn)
(360°)

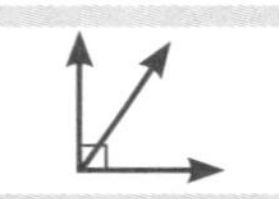

Complementary Angles
Two angles whose measures have a sum of 90°.

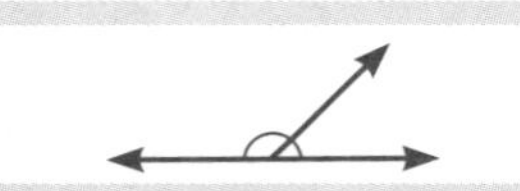

Supplementary Angles
Two angles whose measures have a sum of 180°.

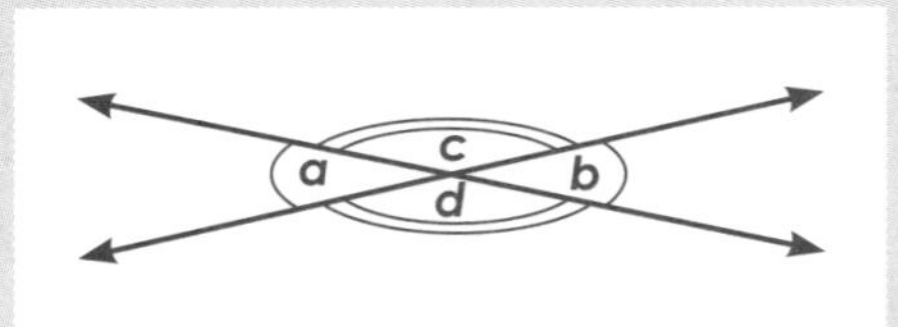

Vertical Angles
Angle pairs a & b, and c & d are vertical and congruent.

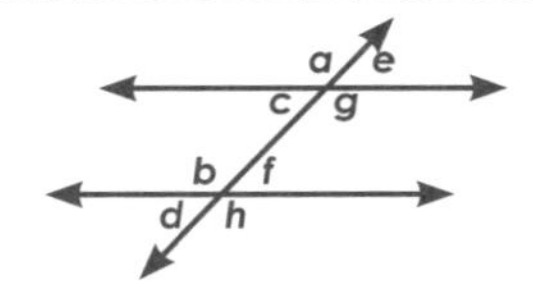

Corresponding Angles
For parallel lines cut by a transversal, pairs a & b, c & d, e & f, g & h are corresponding, congruent angles.

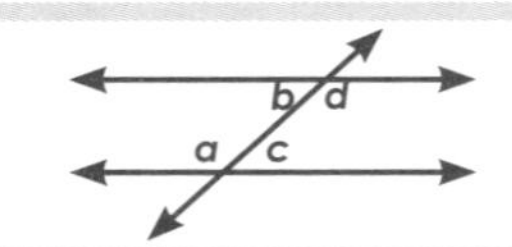

Alternate Interior Angles
For parallel lines cut by a transversal, angle pairs a & d, and b & c are alternate interior angles and are congruent.

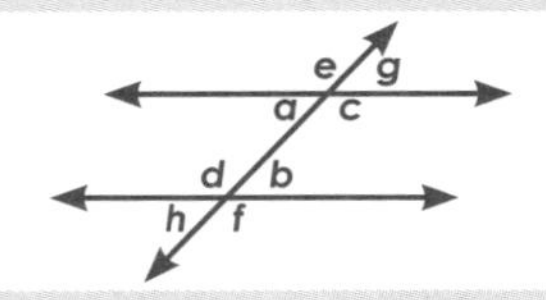

Alternate Exterior Angles
For parallel lines cut by a transversal, angle pairs e & f and g & h are alternate exterior angles and are congruent.

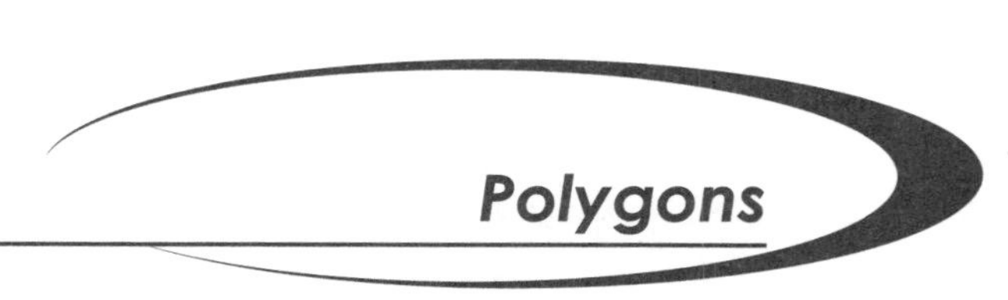

Polygons

A **polygon** is a simple closed plane curve consisting only of line segments. Triangles and quadrilaterals are the most common polygons. [From the Greek *polus* meaning "many" and *gonia* meaning "angle."]

Name	Sides	Irregular	Regular
Triangle	**3**		
Quadrilateral	**4**		
Pentagon	**5**		
Hexagon	**6**		
Heptagon (Septagon)	**7**		
Octagon	**8**		
Nonagon	**9**		
Decagon	**10**		
Undecagon (Unidecagon)	**11**		
Dodecagon (Duodecagon)	**12**		

Triangles

Classifying Triangles

All triangles may be classified according to their sides or according to their angles as follows.

By Sides

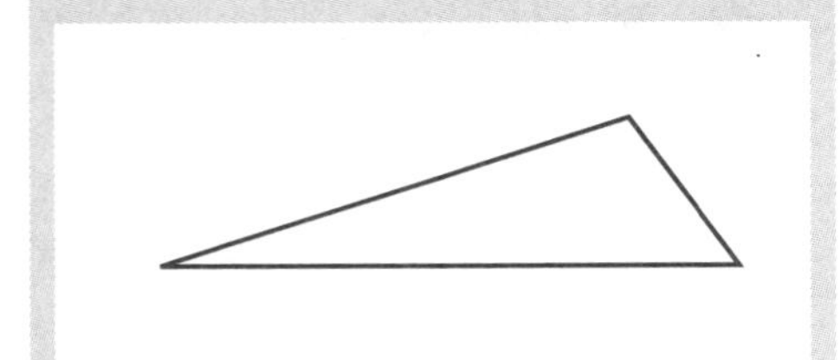

Scalene triangle
No sides (or angles) congruent.

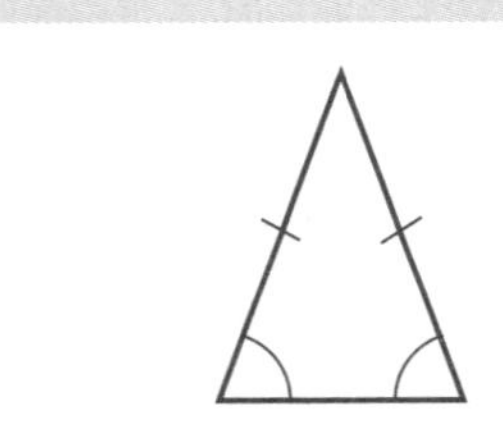

Isosceles triangle
At least two sides congruent. The angles opposite them are also congruent and called base angles.

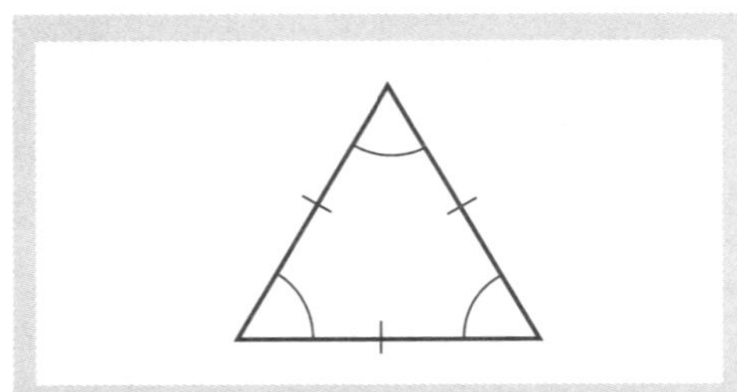

Equilateral triangle
All three sides are congruent. The angles are also congruent (60°).

By Angles

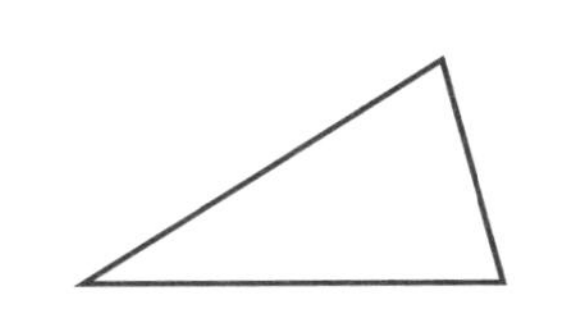

Acute triangle
All angles are acute.

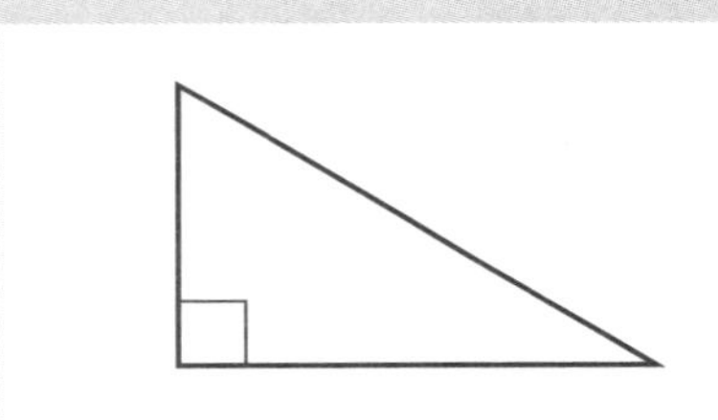

Right triangle
One of the angles is a right angle.

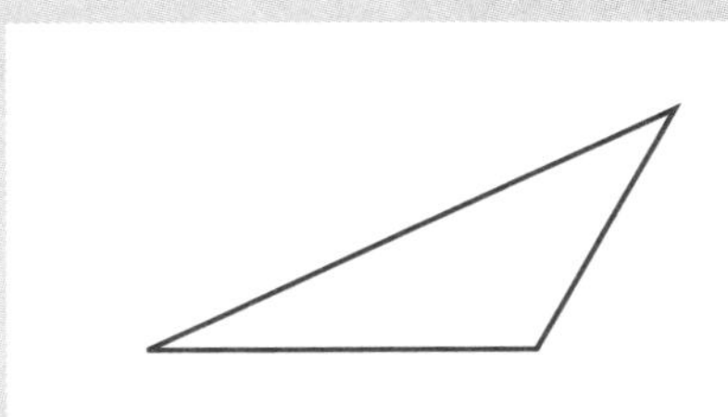

Obtuse triangle
One of the angles is an obtuse angle.

Triangles may also be classified by combining sizes of angles and sides.

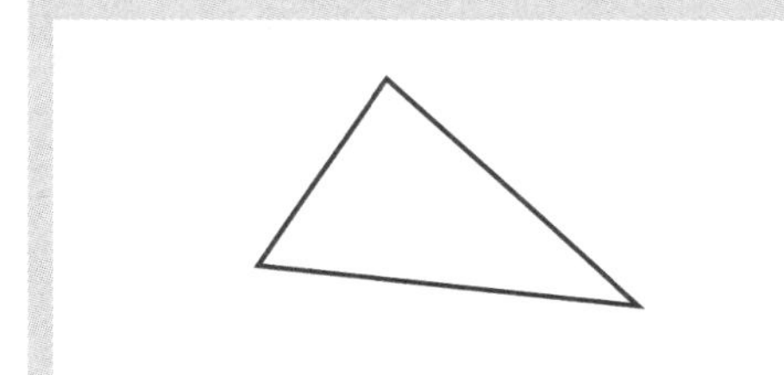

Acute Scalene Triangle
All angles less than 90° and no sides the same length.

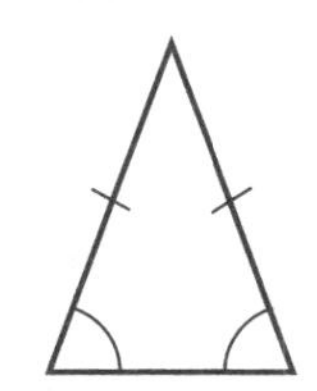

Acute Isosceles Triangle
All angles less than 90° and two sides the same length.

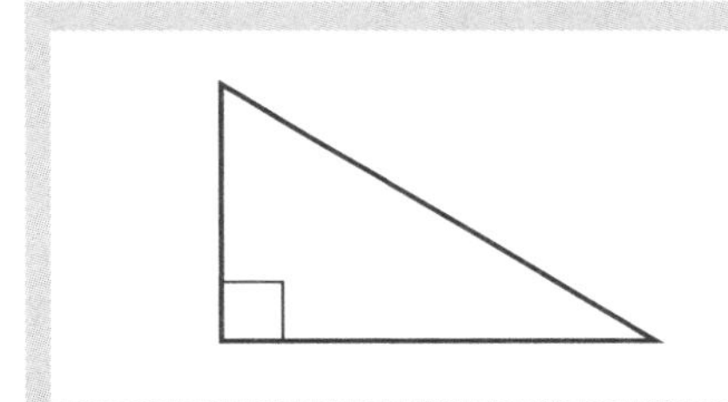

Right Scalene Triangle
One right angle and no sides the same length.

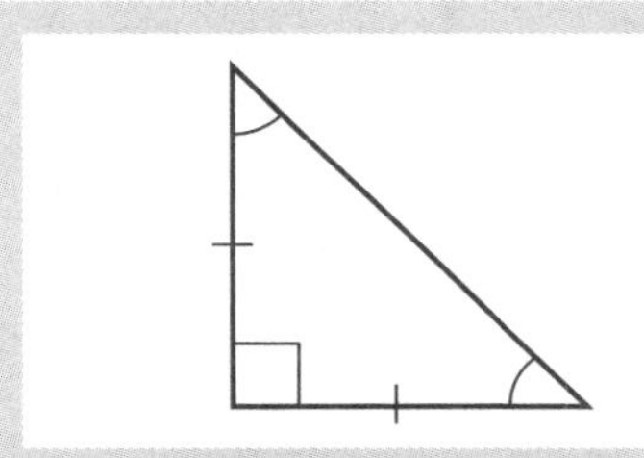

Right Isosceles Triangle
One right angle and two sides the same length.

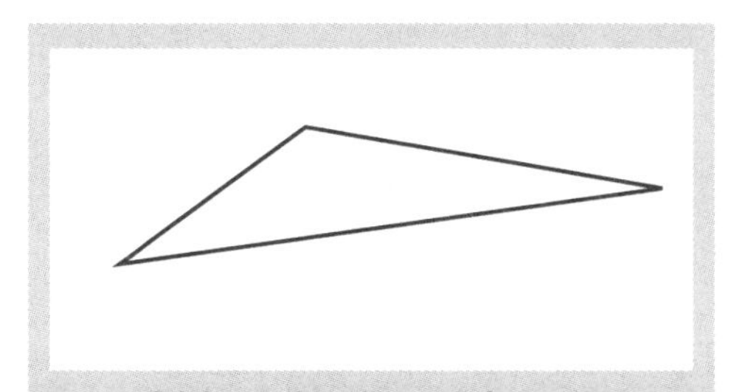

Obtuse Scalene Triangle
One angle greater than 90° and no sides the same length.

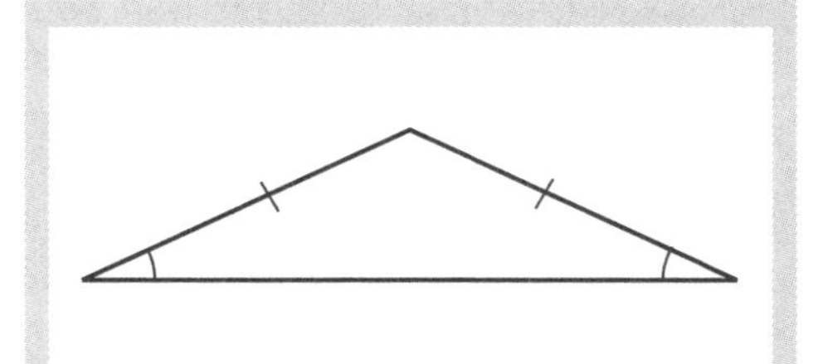

Obtuse Isosceles Triangle
One angle greater than 90° and two sides the same length.

Triangles

Other Triangle Facts

Sum of the Angles in a Triangle

The sum of the angles in a triangle is always 180°.

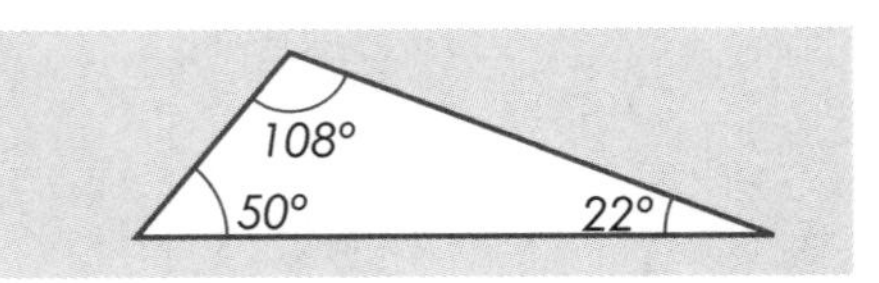

Exterior Angles of a Triangle

Each exterior angle is equal to the sum of the two opposite interior angles.

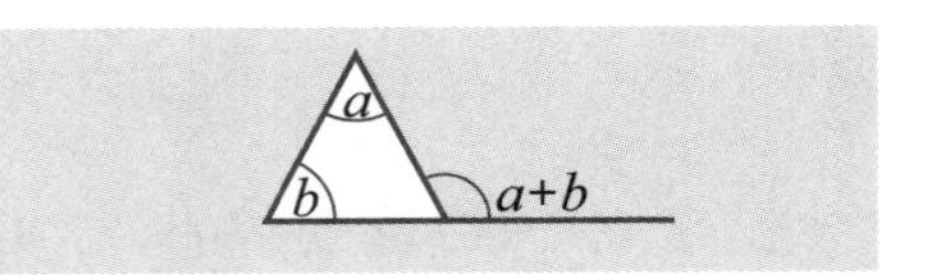

Pythagorean Theorem

For any right triangle, with legs of length *a* and *b* and hypotenuse *c*, the sum of the squares of the legs equals the square of the hypotenuse.

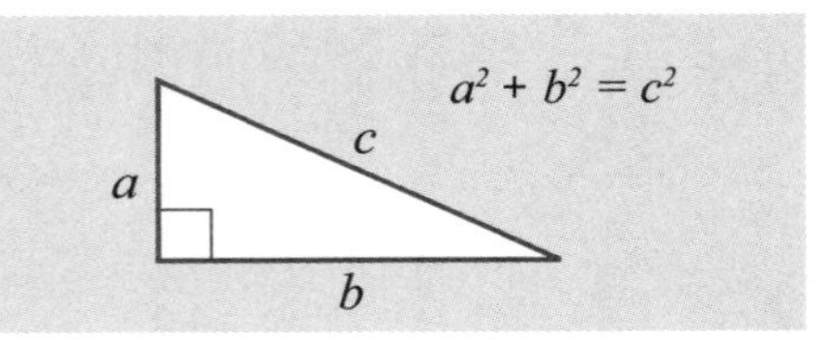

Similar Triangles

Two triangles are similar if they have the same shape, but not necessarily the same size. Being the same shape means that corresponding angles are congruent and the corresponding sides are in the same ratio [see below].

Conditions for Similarity	Diagram
Side-side-side (SSS) *If the lengths of the three corresponding sides (SSS) of two triangles are in the same ratio, then the triangles are similar.*	b c a B C A
Angle-angle (AA) *If two angles of a triangle are congruent to two angles of another triangle, then the triangles are similar.*	
Side-angle-side (SAS) *If one angle of a triangle is congruent to one angle of another triangle and the lengths of the sides (SAS) that determine these angles are in the same ratio, then the triangles are similar. [Note that the angles that are congruent must be between the sides that are in the same ratio, hence the A is placed in the center of the SAS as a reminder.]*	a b A B

Congruent Triangles

Two triangles are congruent if they are exactly the same size and shape. There are several ways to establish that two triangles are congruent.

Conditions for Congruency	Diagram
Side-side-side (SSS) *If three sides of one triangle are congruent to three sides of another triangle, then the triangles are congruent.*	
Side-angle-side (SAS) *If two sides and the included angle of one triangle are the same as two sides and the included angle of another triangle, then the triangles are congruent.*	
Angle-side-angle (ASA) *If two angles and the included side of one triangle are congruent to two angles and the corresponding side of another triangle, then the triangles are congruent.*	
Hypotenuse-angle (HA) *If the hypotenuse and one of the acute angles of a right triangle are congruent to the corresponding hypotenuse and acute angle of the other right triangle, then the triangles are congruent.*	

Trigonometric Ratios

In trigonometry, the sides of a right triangle are given special names. In the triangle ABC, when using angle A as a reference, AB is referred to as the adjacent side, BC as the opposite side and AC as the hypotenuse.

The following trigonometric ratios are defined for angle A.

The sine of angle A is calculated by the ratio:

$$\sin A = \frac{\text{opposite}}{\text{hypotenuse}} = \frac{BC}{AC}$$

The cosine of angle A is calculated by the ratio:

$$\cos A = \frac{\text{adjacent}}{\text{hypotenuse}} = \frac{AB}{AC}$$

The tangent of angle A is calculated by the ratio:

$$\tan A = \frac{\text{opposite}}{\text{adjacent}} = \frac{BC}{AB}$$

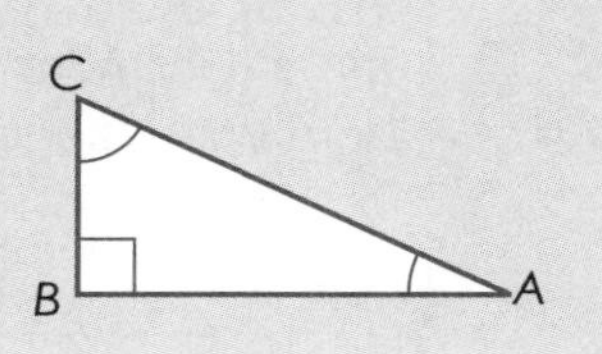

Quadrilaterals

Quadrilaterals are the most common type of polygon in our environment. The different types of quadrilaterals are explained and illustrated in the table below.

Definition	Diagram
Quadrilateral Any polygon that has four sides [Latin *quadri*, four, and *latus*, side.]	
Trapezoid A quadrilateral with at least one pair of parallel sides. [Older mathematics texts define a trapezoid as a quadrilateral with exactly one pair of parallel sides.]	
Isosceles trapezoid A trapezoid with exactly one pair of congruent non-adjacent sides.	
Parallelogram A quadrilateral with both pairs of opposite sides parallel. Properties Sides: Opposite sides are congruent. Angles: Opposite angles are congruent; adjacent angles are supplementary. Diagonals: Triangles formed by each diagonal are congruent; diagonals bisect each other.	
Rhombus A parallelogram with four congruent sides. Properties A rhombus has all the properties of a parallelogram and the following: Diagonals: Diagonals bisect each other at right angles, and also bisect the angles of the rhombus.	

Definition	Diagram
Rectangle A parallelogram with all angles right angles. Properties A rectangle has all the properties of a parallelogram and the following: Diagonals: Its diagonals are the same length.	
Square A rectangle with four congruent sides. Properties A square has all the properties of a parallelogram/rhombus/rectangle.	
Kite A quadrilateral with two distinct pairs of adjacent sides congruent. Properties Diagonals: Its diagonals are perpendicular, and the shorter one is bisected.	

Circles

A **circle** is the set of all points in a plane that are the same distance from a center point. All circles have the same properties; they only vary in size. The components and properties are shown in the table below.

Definition	Diagram
Chord A line segment joining two points of a circle.	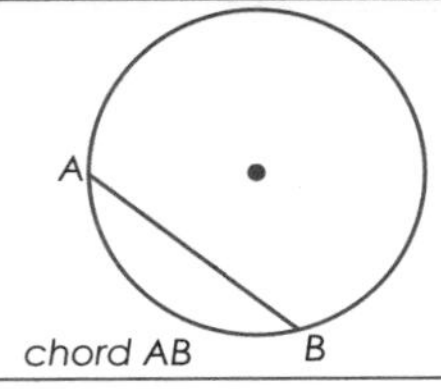 A, chord AB, B
Radius A line segment from the center of the circle to the circle itself, as shown.	 center, radius
Diameter A chord that passes through the center of a circle.	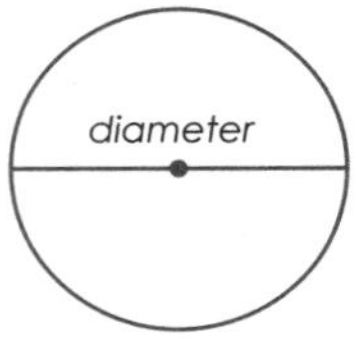 diameter
Circumference The perimeter of or distance around a circle.	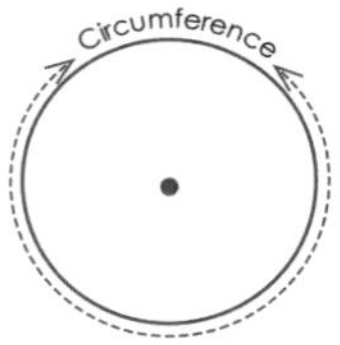 Circumference
Major/Minor Arc A part of a circle between two points on a circle. If the part represents more than half of the circumference, then it is referred to as a major arc; if less, then it is referred to as a minor arc.	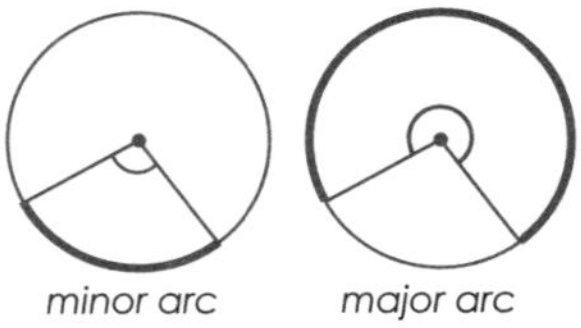 minor arc, major arc
Semicircle Half a circle; any diameter divides a circle into two semicircles.	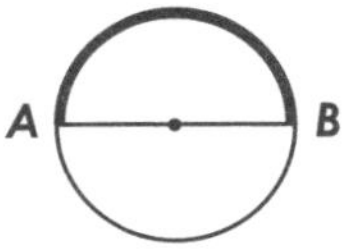 A, B
Circular Region A circle and its interior points.	

Definition	Diagram
Quadrant One fourth of a circular region, formed when two radii are at right angles to each other.	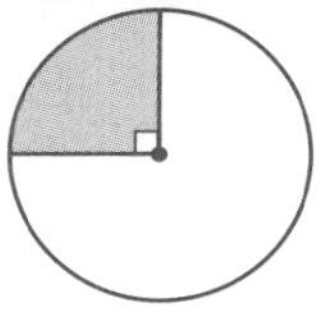
Sector Portion of a circular region bounded by two radii and an arc.	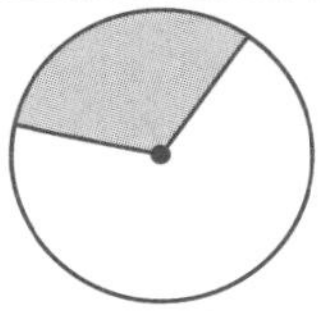
Concentric Circles Two or more circles that share the same center.	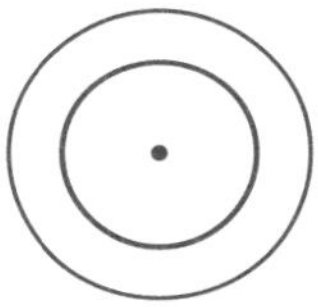
Annulus The plane region between two concentric circles	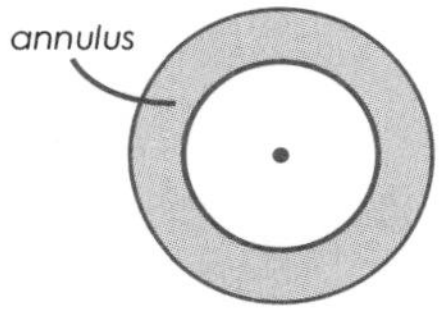 annulus
Tangent (Tangent Line) A line that intersects a circle at exactly one point. A tangent is always perpendicular to the radius at the point of intersection.	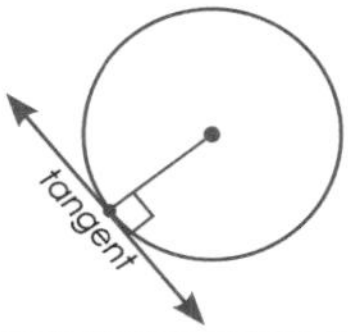 tangent
Circumscribed Circle A circle that passes through all the vertices of a polygon is circumscribed about the polygon.	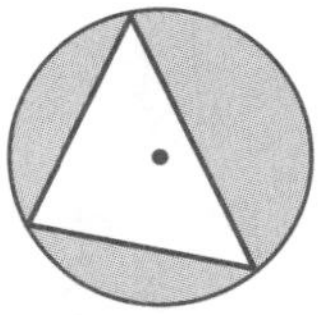
Inscribed Circle A circle that is tangent to all sides of a polygon is inscribed in the polygon.	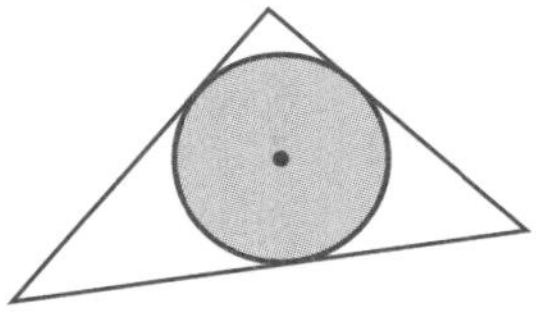

Circles

Circle Angle Properties

Angles in a Segment

All angles in the same segment are congruent.

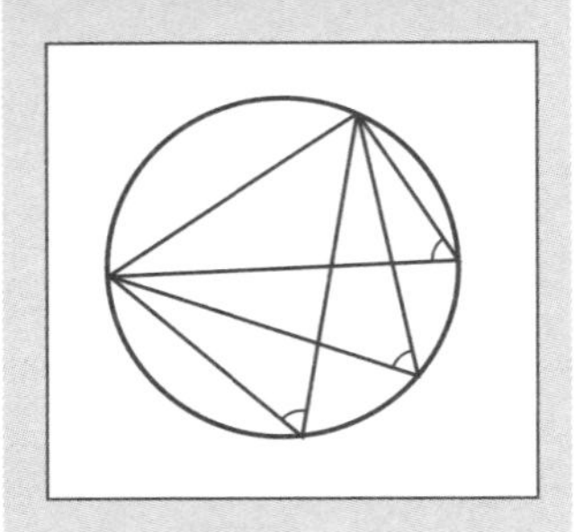

Angles in a Semicircle

All angles in a semicircle are right angles.

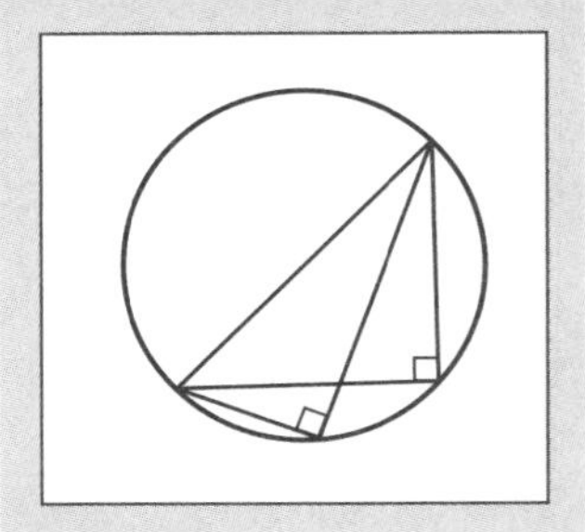

Cyclic Quadrilateral

Opposite angles of a cyclic quadrilateral add up to 180°; $a + c = 180°$; $b + d = 180°$.

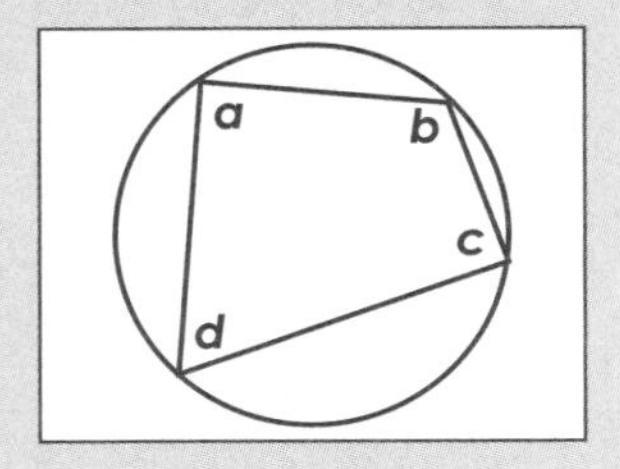

Circle-Related Formulas

Circumference

The circumference (C) of a circle may be calculated by multiplying the diameter (*d*) of a circle by the ratio pi (approx 3.1416) as follows:

$C = \pi \times d$

Alternatively, the diameter is twice the radius ($d = 2r$) so the circumference may be calculated as follows:

$C = \pi \times 2r = 2\pi r$

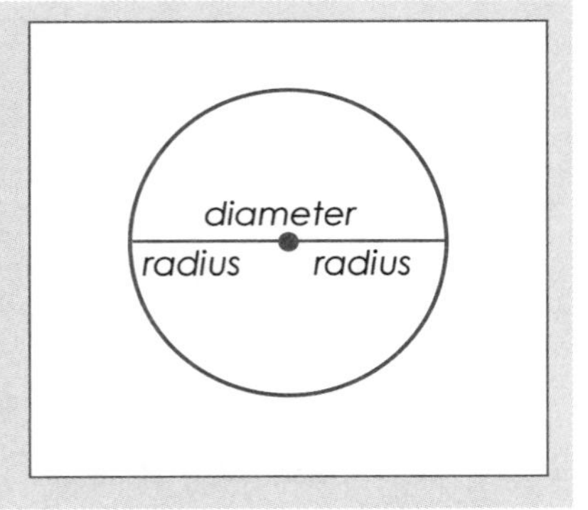

Area of a Circular Region

The area (A) of any circular region can be found as follows, using its radius (r) and the pi ratio.

$A = \pi r^2$

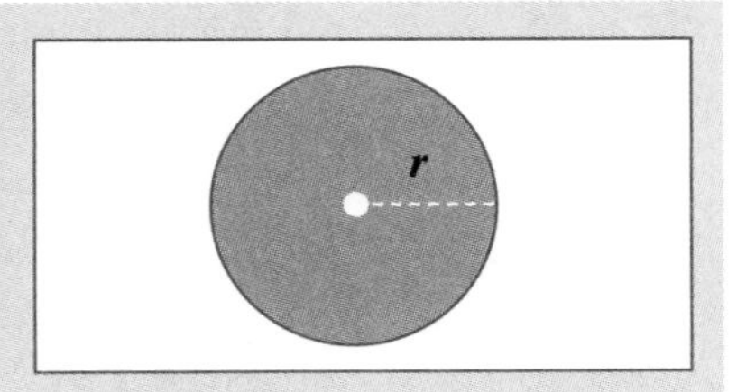

Polyhedra

Polyhedra are three-dimensional (3-D) shapes formed by polygonal regions (faces). The single term is polyhedron.

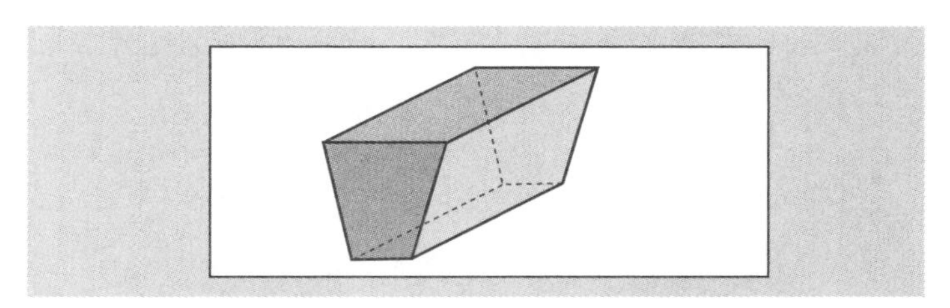

The polyhedron shown here is a hexahedron; i.e. it has six faces.

Prisms

Three-dimensional shapes formed by two congruent polygonal regions in parallel planes (bases), connected by parallelogram regions.

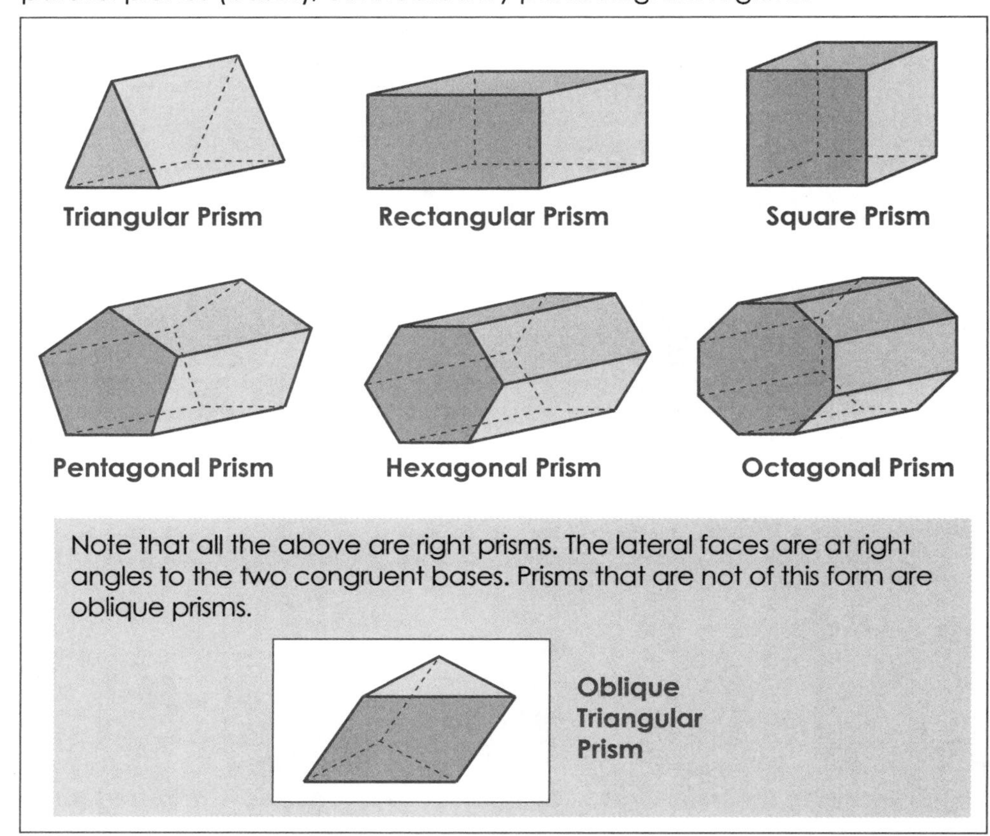

Note that all the above are right prisms. The lateral faces are at right angles to the two congruent bases. Prisms that are not of this form are oblique prisms.

Three-dimensional Shapes

Pyramids

A three-dimensional shape formed by a polygonal region (base), a point (apex) not in the region, and the triangular regions connecting the point to the polygonal region.

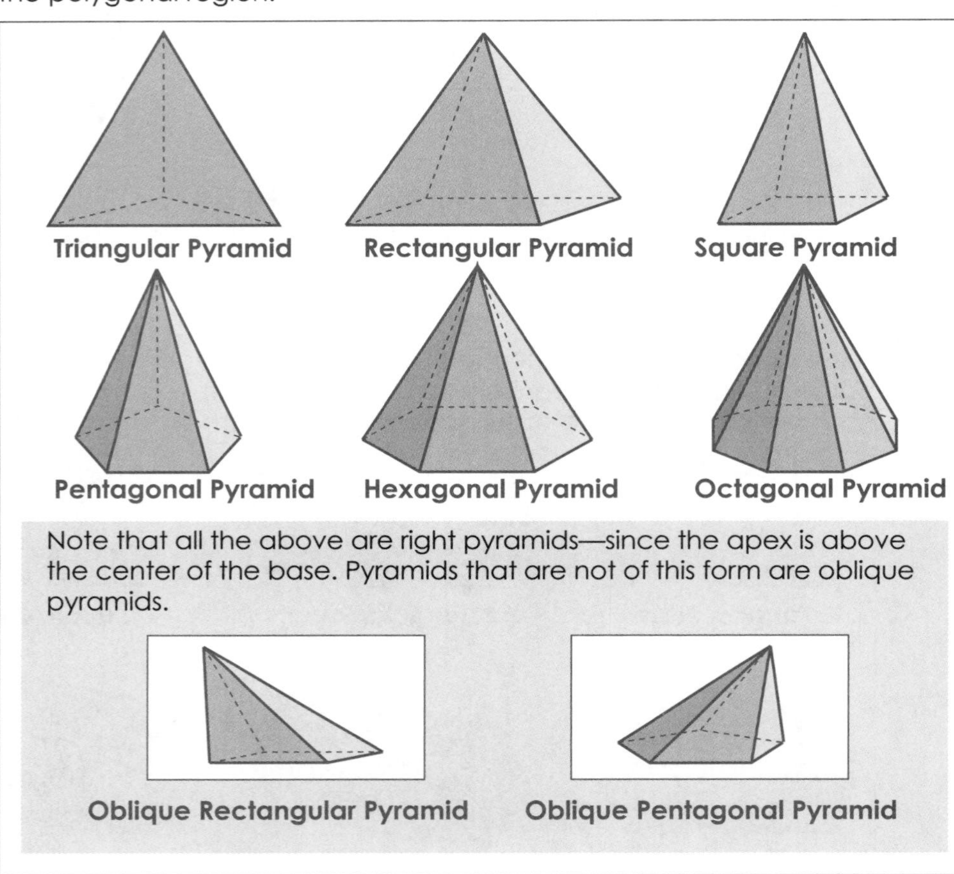

Spheres and Ovoids

There are many three-dimensional shapes that are not polyhedra.

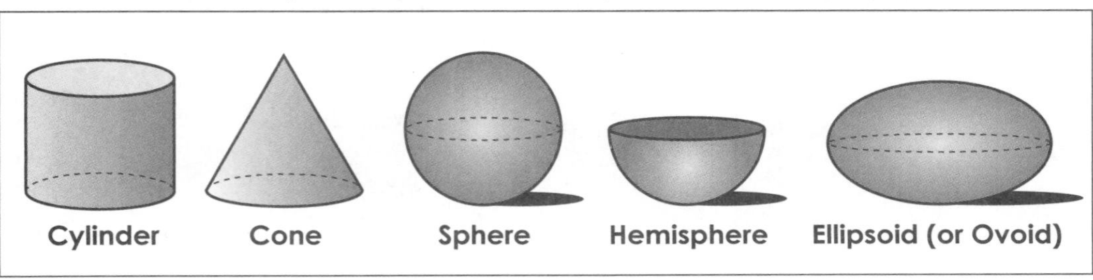

Platonic Solids

The Platonic solids are polyhedra in which all faces are congruent and all interior angles are congruent. There are only five such regular solids. These are illustrated below, together with examples of 2-D nets that fold up to make the 3-D shapes. The 4 faces of the tetrahedron, 8 faces of the octahedron, and 20 faces of the icosahedron are all equilateral triangles. The 6 faces of the hexahedron (cube) are all regular quadrilaterals (squares), while the 12 faces of the dodecahedron are regular pentagons.

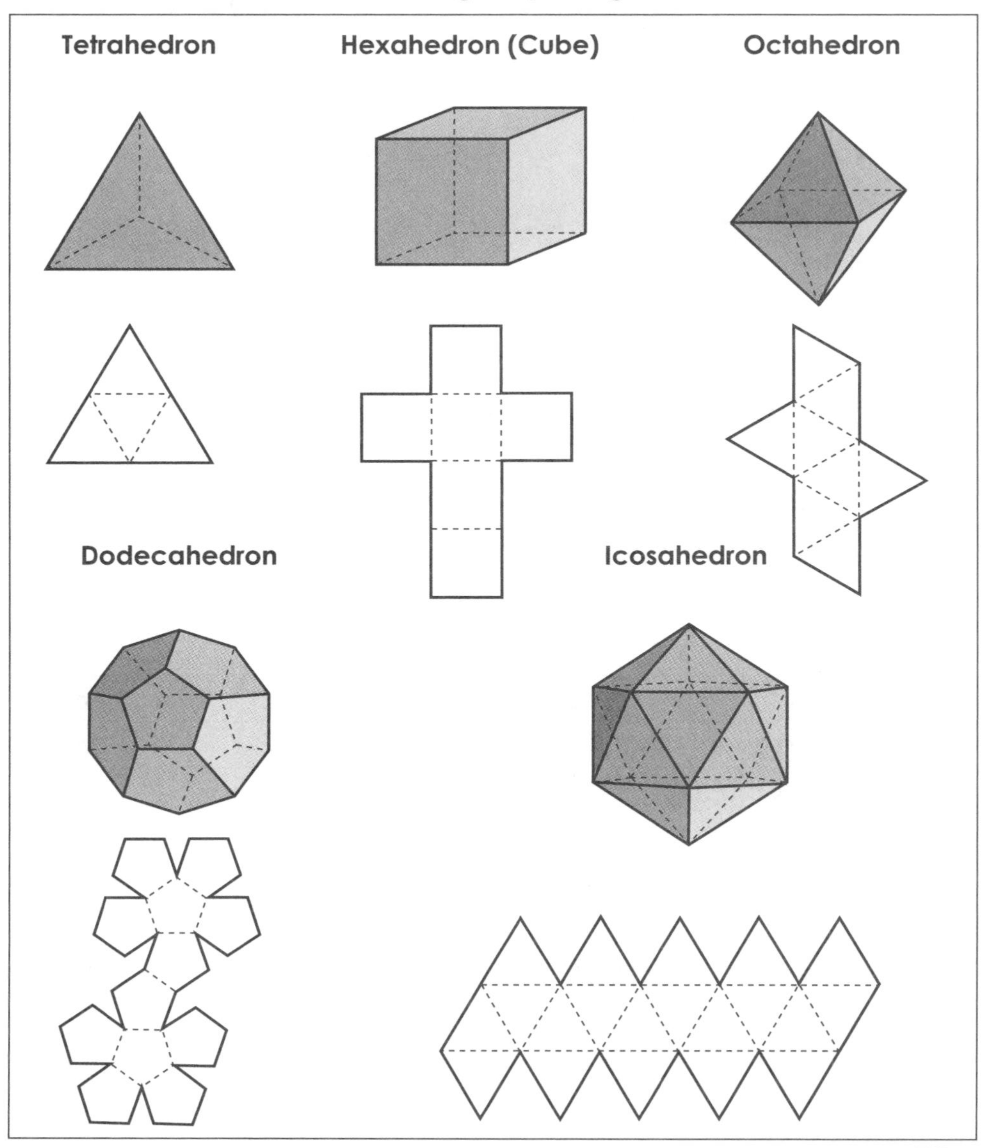

Three-dimensional Shapes

Euler's Formula for Polyhedra

In all polyhedra, there is a relationship between the numbers of vertices (V), faces (F) and edges (E). This relationship can be expressed as $V + F = E + 2$ or $V + F - 2 = E$.

In the square pyramid to the right, there are 5 vertices, 5 faces and 8 edges, so the relationship is seen as $5 + 5 = 8 + 2$.

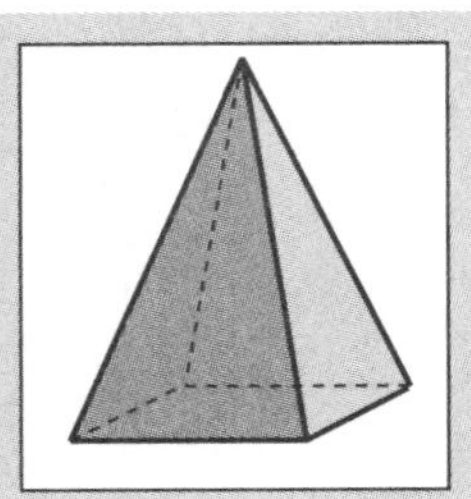

Faces, Edges and Vertices of Polyhedra

Name of Shape	Vertices	Faces	Edges
Triangular Prism	6	5	9
Rectangular Prism	8	6	12
Pentagonal Prism	10	7	15
Hexagonal Prism	12	8	18
Heptagonal Prism	14	9	21
Octagonal Prism	16	10	24
n-agonal Prism	$2n$	$n + 2$	$3n$
Triangular Pyramid	4	4	6
Square Pyramid	5	5	8
Pentagonal Pyramid	6	6	10
Hexagonal Pyramid	7	7	12
Heptagonal Pyramid	8	8	14
Octagonal Pyramid	9	9	16
n-agonal Pyramid	$n + 1$	$n + 1$	$2n$

Reflective (Line) Symmetry

A figure has reflective (or line) symmetry if it can be divided by a line, called the line of symmetry, into two parts that are mirror images of each other.

2-D: Lines of Symmetry

One line of symmetry

More than one line of symmetry

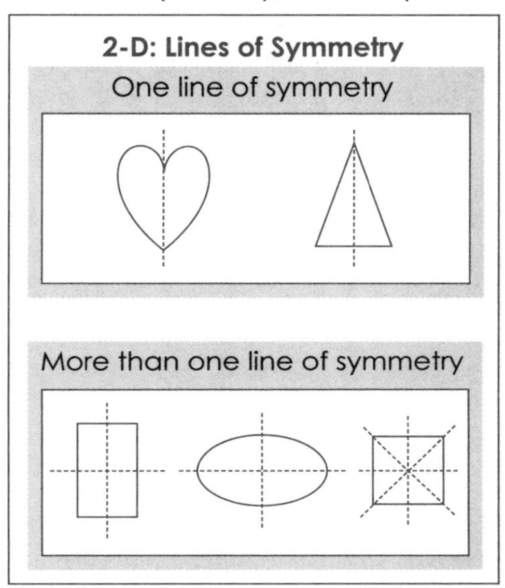

Asymmetry

No lines of symmetry

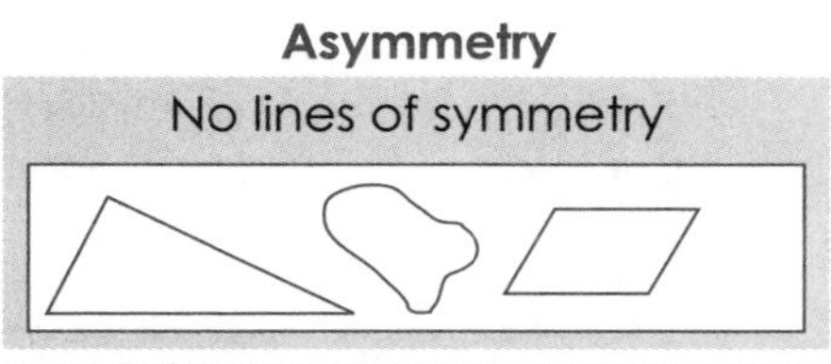

3-D: Planes of Symmetry

The diagram below shows two of the planes of symmetry of a cube, which has a total of nine planes of symmetry.

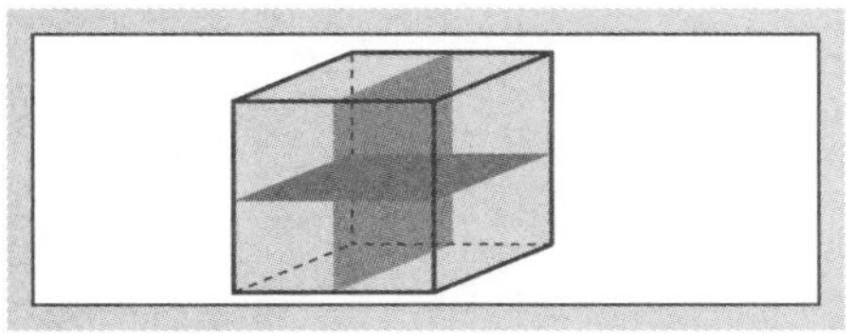

Rotational Symmetry

A figure has rotational symmetry if a turn of 180˚ or less (angle of rotation) about a fixed point (center of rotation) produces an image that fits exactly on the original figure.

This shape has 90° rotational symmetry The image under a 90˚, 180˚, or 270˚ rotation looks the same as the original.

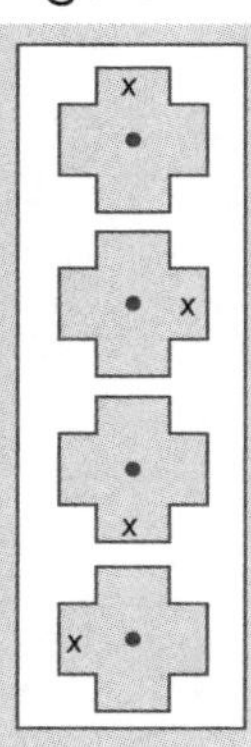

This shape has 120° rotational symmetry The image under a 120˚ or 240˚ rotation looks the same as the original.

If a shape can only be rotated 360˚ (a full turn) to return to its original position, then it does not have rotational symmetry. For example, the triangle below does not have rotational symmetry.

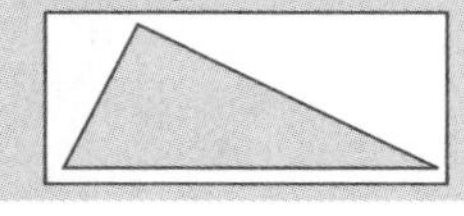

3-D: Axes of Symmetry

The diagram below shows one of the axes of symmetry of a regular tetrahedron. It has 120˚ rotational symmetry about the axis shown.

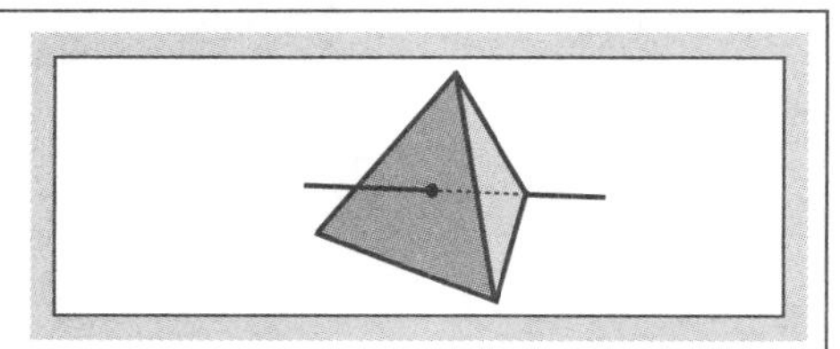

Transformations

A transformation is the process by which figure is changed in shape, size, or position.

Isometries

Translations, reflections and rotations are all isometries. In these transformations, length, width, angle size and area do not change. Thus, the original figure (preimage) and its image are congruent.

Translation (Slide)

A transformation that moves each point of a figure the same distance in the same direction.

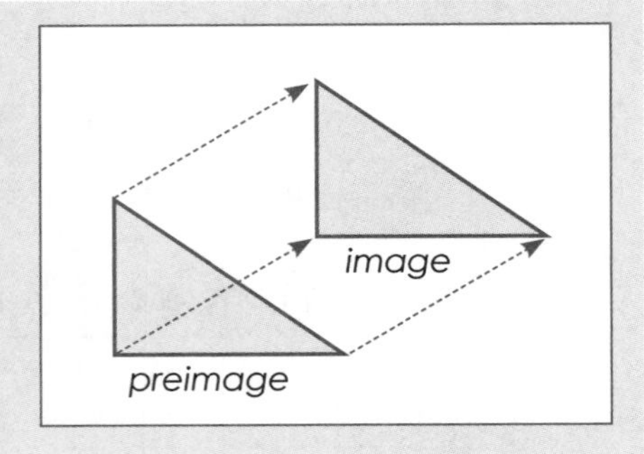

Reflection (Flip)

A transformation that reflects a figure in (or about) a line, called the line of reflection, creating a mirror image of the figure.

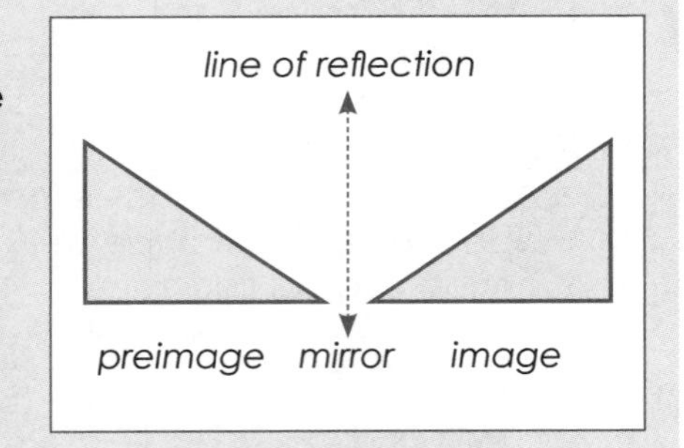

Rotation (Turn)

A transformation that turns a figure in a given direction through a given angle (angle of rotation) about a fixed point (center of rotation).

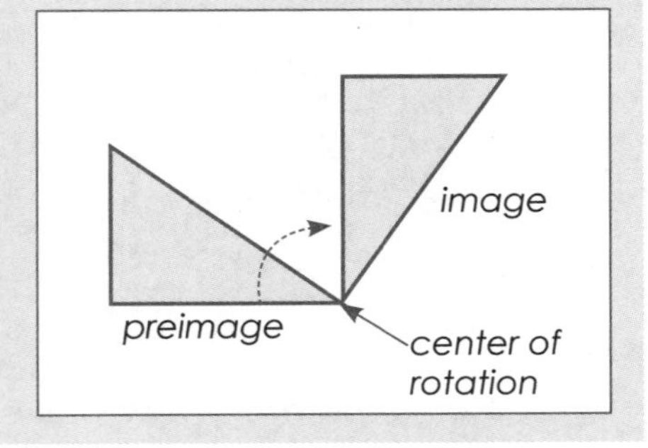

Size Transformations

A size transformation is a transformation that expands (stretches) or contracts (shrinks) a figure. Size transformations are not isometries, though the resulting image is the same shape as (similar to) the original.

Expansion

An expansion is an increase in the size of a figure while maintaining the same shape.

Expansion about a point

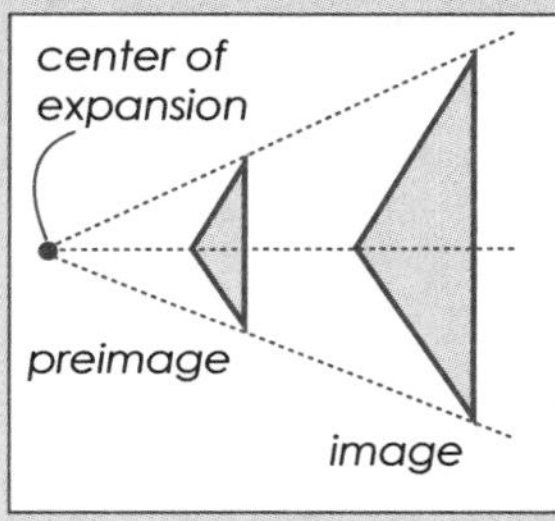

Expansion by grid

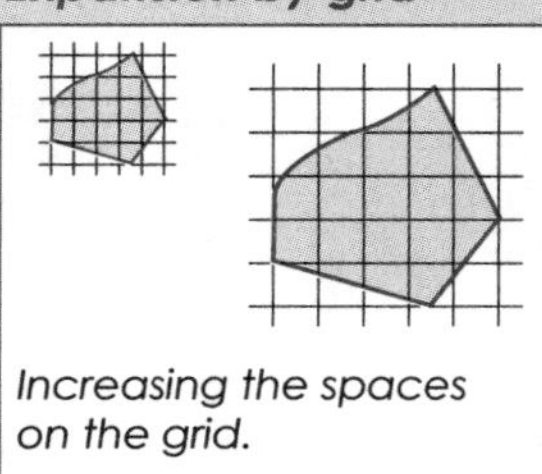

Increasing the spaces on the grid.

Contraction

A contraction is a decrease in the size of a figure while maintaining the same shape.

Contraction about a point

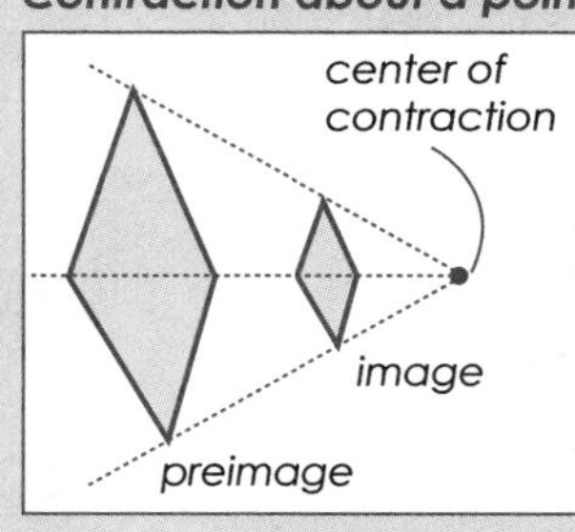

Contraction by grid

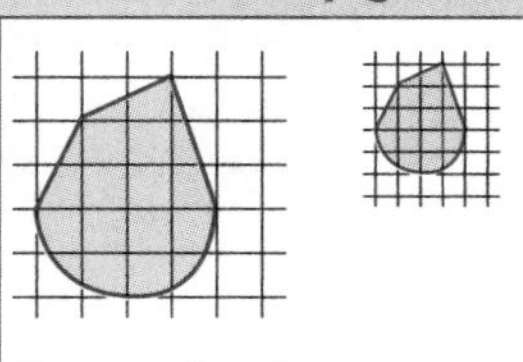

Decreasing the spaces on the grid.

Distortion Transformations

A distortion transformation is a transformation that changes both the shape and size of a figure.

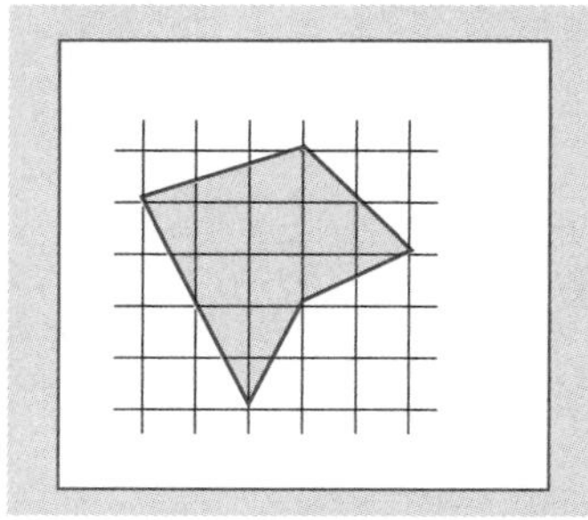

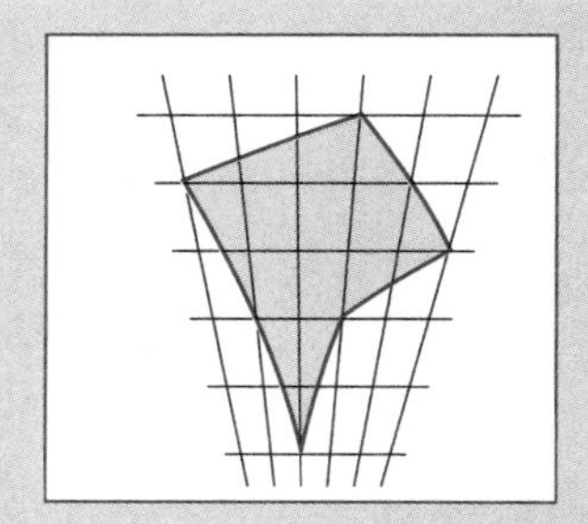

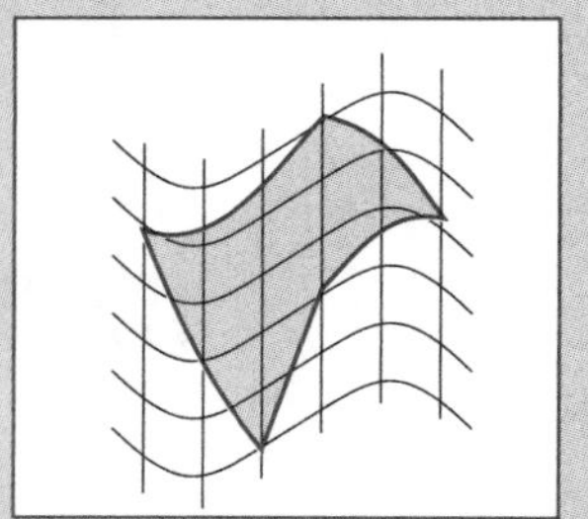

Cross-Sections and Conic Sections

Cross-sections refer to the plane regions resulting from planar cuts through 3-D objects.

Some Cross-Sections of Cubes

Cutting a cube horizontally or vertically parallel to any face produces a square cross-section.

Cutting a cube from one edge to another produces a rectangular cross-section.

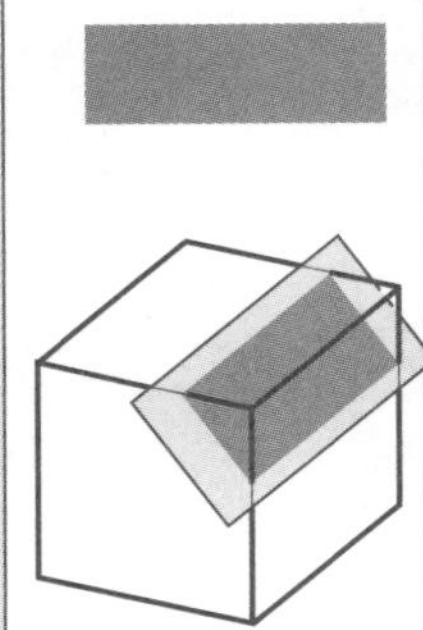

Cutting an edge off a cube produces a rectangular cross-section.

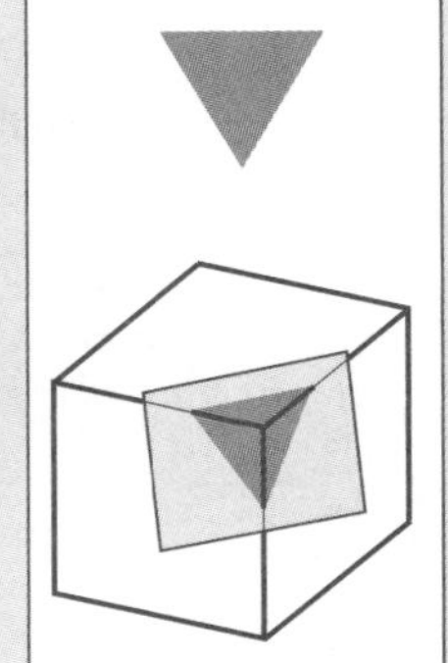

Cutting a corner off a cube produces a triangular cross-section.

There are many other resulting cross sections of a cube, such as trapezoids and hexagons.

Some Cross-Sections of Cylinders

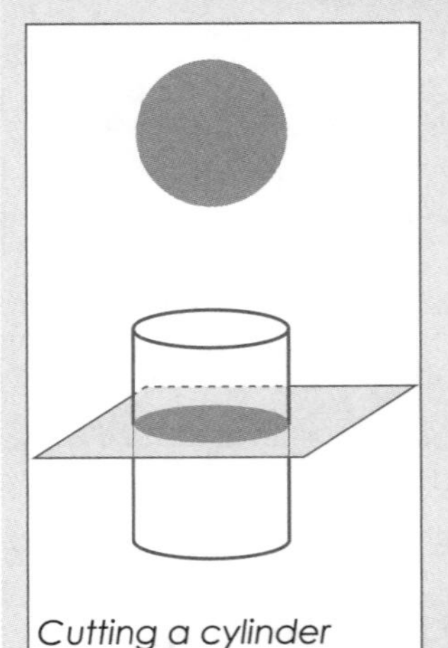

Cutting a cylinder horizontally produces a circular cross-section.

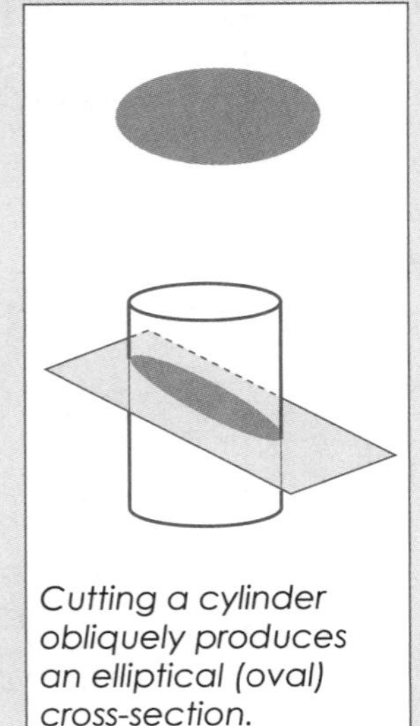

Cutting a cylinder obliquely produces an elliptical (oval) cross-section.

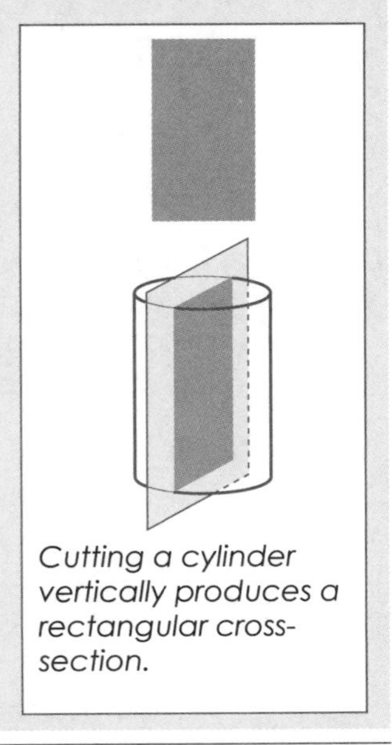

Cutting a cylinder vertically produces a rectangular cross-section.

Cross-Sections and Conic Sections

Cross-Sections of Cones (Conic Sections)

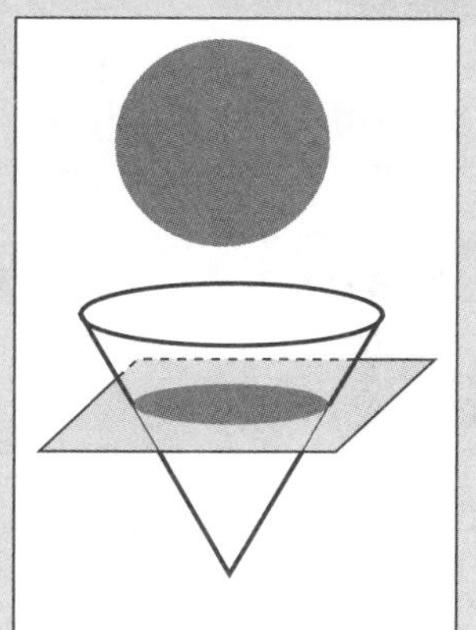

Cutting a cone horizontally produces a circular cross-section.

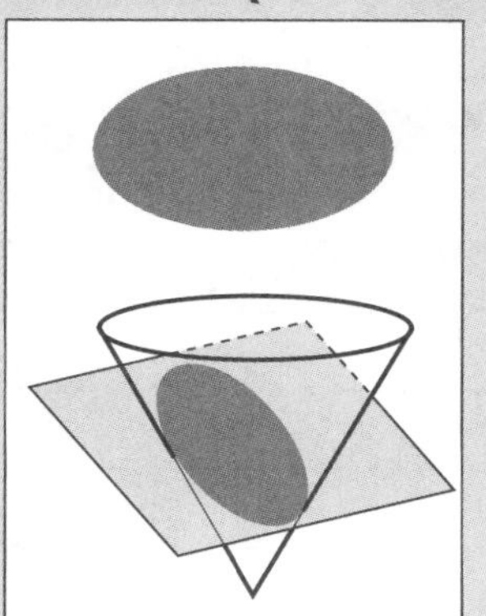

Cutting a cone obliquely produces an elliptical (oval) cross-section.

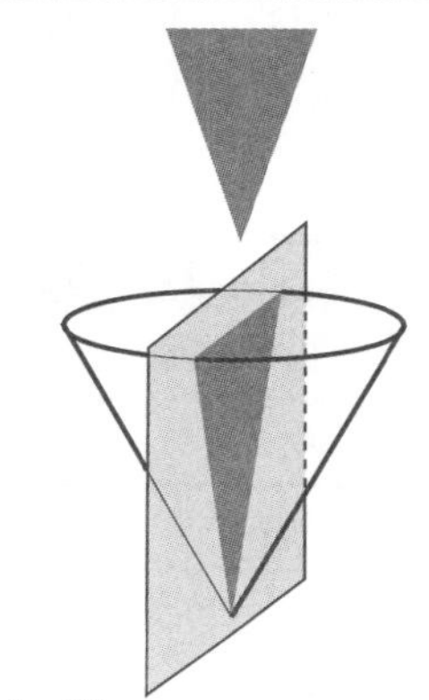

Cutting a cone vertically through the center produces an isosceles triangular cross-section.

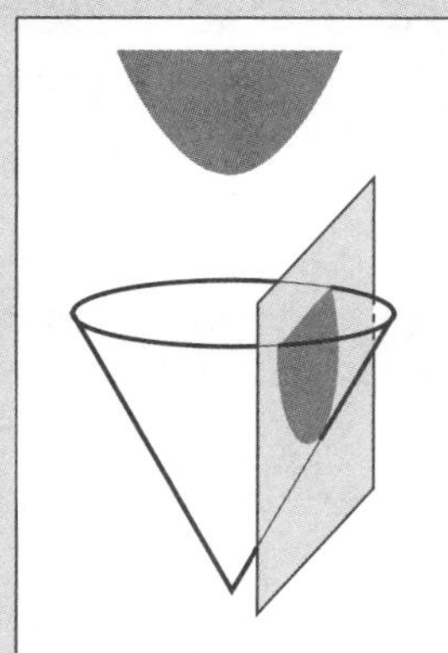

Cutting a cone vertically but not through the center produces half a hyperbola cross-section.

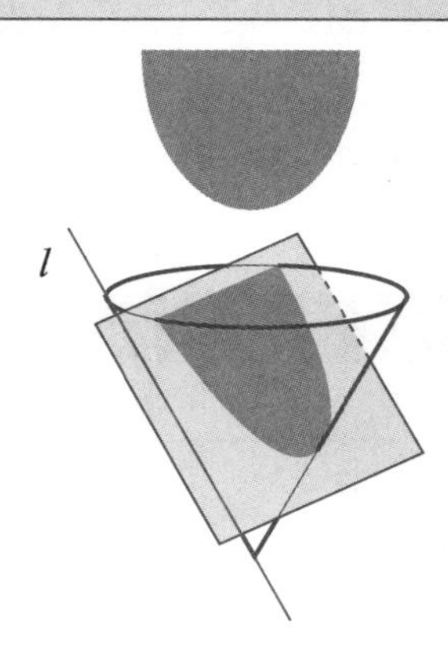

Cutting a cone diagonally and parallel to line l on its curved surface produces a parabola cross-section.

The Metric System and the International System (SI)

There are three kinds of SI units:

1. Base units
2. Supplementary units
3. Derived units—Complex names and Special names

SI Base Units

Physical Quantity	Unit	Symbol
Length	meter	m
Mass	kilogram	kg
Time	second	s
Electric current	ampere	A
Thermodynamic temperature	kelvin	K
Luminous intensity	candela	cd
Amount of substance	mole	mol

SI Supplementary Units

Physical Quantity	Unit	Symbol
Plane angle	radian	rad
Solid angle	steradian	sr

Examples of Derived Units—Complex Names

Physical Quantity	Unit	Symbol
Area	square meter	m^2
Volume	cubic meter	m^3
Density	kilogram per cubic meter	kg/m^3
Speed	meter per second	m/s

Examples of Derived Units—Special Names

Physical Quantity	Unit	Symbol
Force	newton	N
Pressure	pascal	Pa
Work/Energy/Heat	joule	J
Power	watt	W

The Metric System and the International System (SI)

In SI there is only one unit name for each physical quantity, with multiples of the unit recorded by using prefixes.

Metric Prefixes and Names

Prefix	Symbol	Power	Name	Numeral
tera	T	10^{12}	trillion	1,000,000,000,000
giga	G	10^{9}	billion	1,000,000,000
mega	M	10^{6}	million	1,000,000
kilo	k	10^{3}	thousand	1,000
hecto	h	10^{2}	hundred	100
deka	da	10^{1}	ten	10
deci	d	10^{-1}	tenth	0.1
centi	c	10^{-2}	hundredth	0.01
milli	m	10^{-3}	thousandth	0.001
micro	μ	10^{-6}	millionth	0.000001
nano	n	10^{-9}	billionth	0.000000001
pico	p	10^{-12}	trillionth	0.000000000001

SI does not use all the metric system prefixes but only the base units in multiples of thousands and thousandths. For example the most common SI length units used are the meter (base unit, m), a thousandth of a meter (millimeter, mm), and a thousand times a meter (kilometer, km). Thus h, da, d, c are not generally used internationally, but some of these prefixes are in wide use in many countries. For example, the centimeter (cm) is used for school, household and clothing measurements. Below are the common equivalents and approximate factors for some mental conversions.

Length

10 millimeters (mm) = 1 centimeter (cm)
1000 millimeters = 1 meter (m)
100 centimeters = 1 meter
1000 meters = 1 kilometer (km)

Approximate Metric/English equivalent lengths

25 mm ≈ 1 inch
2.5 cm ≈ 1 inch
(so 1 cm ≈ 0.4 inch)
30 cm ≈ 1 foot
91 cm ≈ 1 yard
1.6 km ≈ 1 mile
(so 1 km ≈ 0.6 mile)

The Metric System and the International System (SI)

Area

100 square millimeters (mm^2) = 1 square centimeter (cm^2)
10,000 square centimeters = 1 square meter (m^2)
1,000,000 square meters = 1 square kilometer (km^2)
10,000 square meters = 1 hectare (ha)
100 hectares = 1 square kilometer

Approximate Metric/ English Equivalent Areas

1 ha ≈ 2.5 acres
(so 1 acre ≈ 0.4 ha)
1 km^2 ≈ 0.4 square miles
(so 1 square mile ≈ 2.5 km^2)

Volume

1000 cubic millimeters (mm^3) = 1 cubic centimeter (cm^3)
1000 cubic centimeters = 1 cubic decimeter (dm^3)
1,000,000 cubic centimeters = 1 cubic meter (m^3)

Approximate Metric/English Equivalent Volumes

1 m^3 ≈ 1.3 cubic yards
(so 1 cubic yard ≈ 0.75 m^3)

Capacity

1000 milliliters (mL) = 1 liter (L)
1000 liters = 1 kiloliter (kL)

Approximate Metric/English Equivalent Capacities

1 L ≈ 0.22 gallons
(so 1 gallon ≈ 4.5 L)

Mass

1000 grams (g) = 1 kilogram (kg)
1000 kilograms = 1 tonne (t)

Approximate Metric/English Equivalent Masses

1 kg ≈ 2.2 pounds
(so 1 pound ≈ 0.45 kg)
1t ≈ 1 ton

Speed

1 kilometer per hour (km/h) ≈ 0.28 meters per second (m/s)

Temperature

The degree Celsius (°C) rather than the kelvin is used for everyday situations. On the Celsius scale the freezing point of pure water at sea level is 0° C and the boiling point is 100° C.
Note that:

0° C = 32° Fahrenheit

100° C = 212° Fahrenheit

$$C = \frac{5}{9}(F - 32)$$

$$F = \frac{9}{5}C + 32$$

The Mass/Volume/Capacity Link

1. One cubic centimeter of water is one milliliter and has a mass of one gram; i.e. 1 cm^3 = 1 mL = 1 g of water.
2. One cubic decimeter of water is one liter and has a mass of one kilogram; i.e. 1 dm^3 = 1 L = 1 kg of water.
3. One cubic meter of water is one kiloliter and has a mass of one tonne; i.e. 1 m^3 = 1 kL = 1 t of water.

All countries also use non-SI units, especially those for time and angular measure, and in both these cases there is a link to the ancient Babylonians who used a base of 60 in their numeration system.

Time

60 seconds (s)	= 1 minute (min)
60 minutes	= 1 hour (h)
24 hours	= 1 day
7 days	= 1 week
2 weeks	= 1 fortnight
52 weeks 1 day	= 1 year
52 weeks 2 days	= 1 leap year
365 days	= 1 year
366 days	= 1 leap year
12 months	= 1 year
10 years	= 1 decade
100 years	= 1 century
1000 years	= 1 millennium

24-hour Time (Military time)

Time can be expressed in 12-hour (a.m./p.m.) format or in 24-hour format. Note that when writing 24-hour time, neither punctuation nor spaces are used; e.g. 0638 = 6:38 a.m.; 1245 = 12:45 p.m.; 2217 = 10:17 p.m. The graphic table below can be used to convert from one format to another.

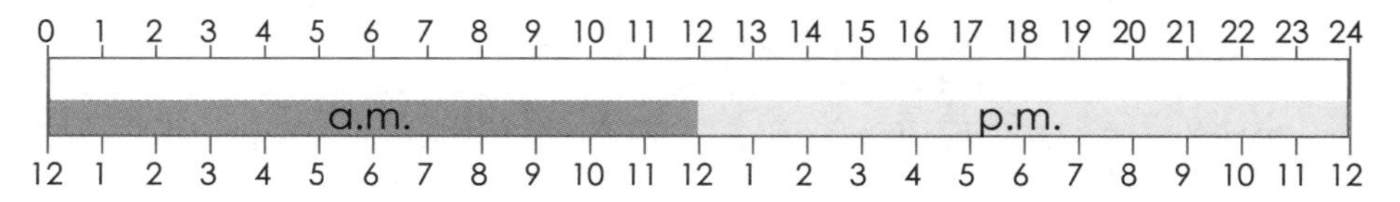

Months of the Year

January	31 days	July	31 days
February	28 or 29 days	August	31 days
March	31 days	September	30 days
April	30 days	October	31 days
May	31 days	November	30 days
June	30 days	December	31 days

Time Units

Days in Each Month

The verses below can be used to recall the days in each month.

Thirty days have September,
April, June and November.
February has twenty-eight;
And thirty-one the others date.
But if a leap year to assign,
Then February twenty-nine.

Thirty days hath September,
April, June and November.
All the rest have thirty-one;
Excepting February alone,
Which has twenty-eight days clear,
And twenty-nine in each leap year.

By clenching both fists alongside each other, the days in all the months can also be recalled by checking the knuckles and the gaps as shown in the diagram.

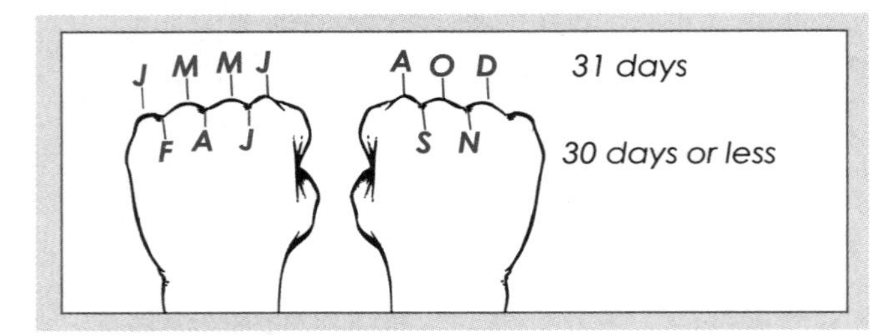

Years and Leap Years

The length of a solar year is almost $365\frac{1}{4}$ days (actually 365 days, 5 hours, 48 minutes, 45 seconds) as this is the time it takes the Earth to complete its orbit around the Sun. Thus, each four years it is necessary to add one day, so that while a year has 365 days, a leap year has 366 days. The years 1980, 1984, 1988, 1992, 1996, and 2004 were all leap years because each number is divisible by four. However, to correct for the fact that a day added every four years is slightly too much, century years are only leap years if divisible by 400. Thus 1600 and 2000 were both leap years, but 1700, 1800 and 1900 were not leap years. [This slight change was made by Pope Gregory in 1582, so the calendar in universal use is called the Gregorian Calendar, although it was originally developed by Julius Caesar, with a slight change made by Caesar Augustus—thus the months of Julius (July) and Augustus (August). The old Roman calendar had 10 months and we still use the Latin prefixes for 7, 8, 9 and 10 in the names for September, October, November and December.]

Angular Measure

60 seconds = 1 minute
60 minutes = 1 degree
90 degrees = 1 right angle
180 degrees = 1 straight angle
360 degrees = 1 rotation
360 degrees = 2π radians
1 radian = $180/\pi \approx 57$ degrees

Example: 36° 40' 6" is an angle of measure 36 degrees, 40 minutes and 6 seconds.

Distance & Speed

The nautical mile will continue to be used for navigational purposes because it is the length of an arc on the Earth's surface formed by an angle of one minute ($\frac{1}{60}°$) at the Earth's center. The knot is a unit of speed of one nautical mile per hour, that is usually applied to wind and boat speeds.

1 nautical mile ≈ 1.85 kilometers
1 nautical mile ≈ 1.16 English miles
1 knot (kn) = 1 nautical mile per hour
1 kn ≈ 1.85 km/h

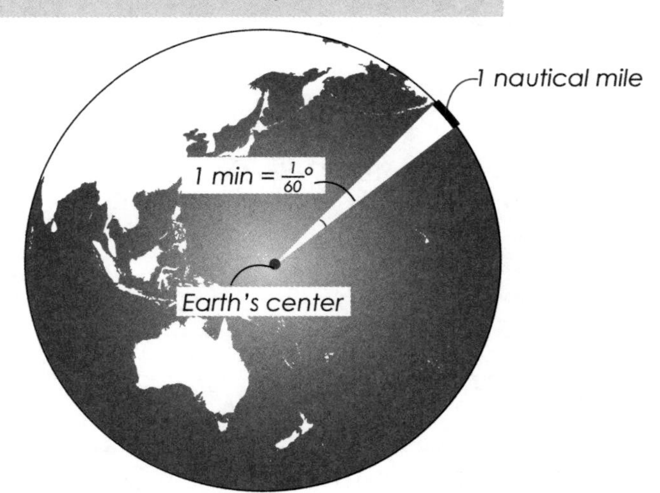

Note: Angle not drawn to scale.

Since a 1° angle at the Earth center corresponds to 60 nautical miles, the circumference of the Earth = 360° x 60 nautical miles = 21,600 nautical miles.

Customary (English) System of Measurement

The Customary (English) system of measurement is still in use in several countries, including the United States.

Units of Length		
Unit	**Abbreviation/Symbol**	**Relationship**
inch (inches)	in.	
foot (feet)	ft.	1 foot = 12 inches
yard	yd.	1 yard = 3 feet = 36 inches
mile	mi.	1 mile = 1760 yards = 5280 feet
Other Units of Length		
rod		1 rod = $5\frac{1}{2}$ yards
furlong		1 furlong = 40 rods
fathom		1 fathom = 6 feet
Units of Area		
Unit	**Abbreviation/Symbol**	**Relationship**
square inch	in.^{2}	
square foot	ft.^{2}	1 ft.^{2} = 144 in.^{2}
square yard	yd.^{2}	1 yd.^{2} = 9 ft.^{2}
acre		43,560 ft.^{2}
square mile	mi.^{2}	640 acres
Units of Volume or Capacity		
Unit	**Abbreviation/Symbol**	**Relationship**
cubic inch	in.^{3}	
cubic foot	ft.^{3}	1 ft.^{3} = 1728 in.^{3}
cubic yard	yd.^{3}	1 yd.^{3} = 27 ft.^{3}
Other Units of Volume or Capacity		
teaspoon	tsp.	
Tablespoon	Tbs.	1 Tbs. = 3 tsp.
cup		1 cup = 16 Tbs.
pint	pt.	1 pt. = 2 cups
quart	qt.	1 qt. = 2 pts.
gallon	gal.	1 gal. = 4 qts. = 8 pts. 1 gal. = 231 in.^{3}
Units of Weight		
Unit	**Abbreviation/Symbol**	**Relationship**
ounce	oz.	
pound	lb.	1 lb. = 16 oz.
ton	T.	1 T. = 2000 lb.

Metric ↔ Customary (English) Conversion Table

The conversion factors listed below are shown to four significant figures and are more precise than those given on the preceding pages. Use a calculator to work particular examples, following these methods:

1. To convert from Customary (English) measures to Metric or SI measures, **multiply** by the conversion factor.
2. To convert from Metric or SI measures to Customary (English) measures, **divide** by the conversion factor.

Quantity	Customary (English)	Conversion Factor	Metric
Length	inches	25.40	millimeters
	inches	2.540	centimeters
	feet	0.3048	meters
	yards	0.9144	meters
	miles	1.609	kilometers
Area	square inches	645.2	square millimeters
	square inches	6.452	square centimeters
	square feet	0.0929	square meters
	square yards	0.8361	square meters
	acres	0.4047	hectares
	square miles	2.590	square kilometers
Volume	cubic inches	16 390	cubic millimeters
	cubic inches	16.39	cubic centimeters
	cubic feet	0.0283	cubic meters
	cubic yards	0.7646	cubic meters
Capacity	fluid ounces	28.41	milliliters
	pints	0.5683	liters
	gallons	4.546	liters
Mass	ounces	28.35	grams
	pounds	0.4536	kilograms
	tons	1.016	tonnes
Speed	feet/second	0.3048	meters/second
	feet/minute	0.005080	meters/second
	miles/hour	1.609	kilometers/hour
	miles/hour	0.4470	meters/second

Rules for Finding Area

Square Region

Area = side x side

A = s^2

Rectangular Region

Area = length x width (or base x height)

A = lw (or bh)

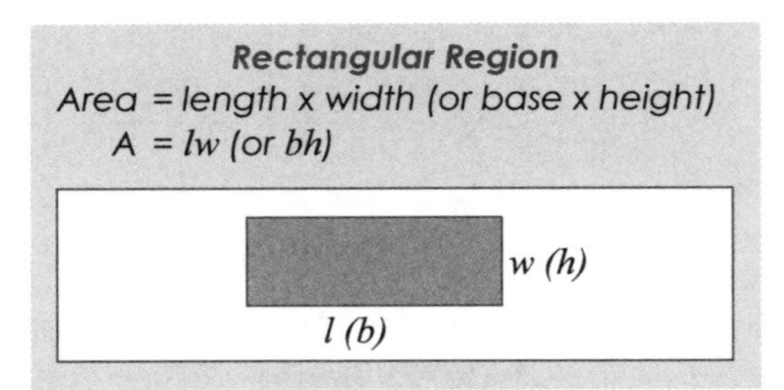

Parallelogram and Rhombus Regions

Area = base x height

A = bh

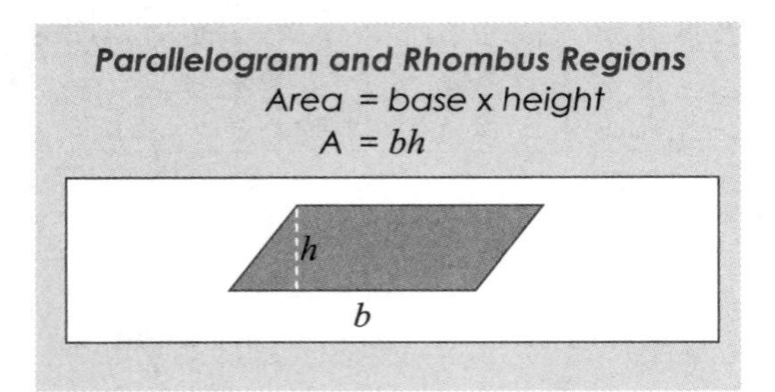

Trapezoidal Region

Area = $^1/_2$ x sum of parallel sides lengths x height

A = $^1/_2\,(a + b)h$

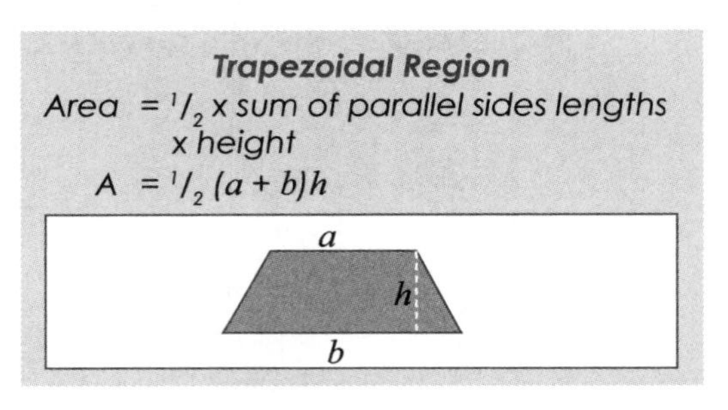

Triangular Region

Area = $\frac{1}{2}$ x base x height

A = $\frac{1}{2}bh$

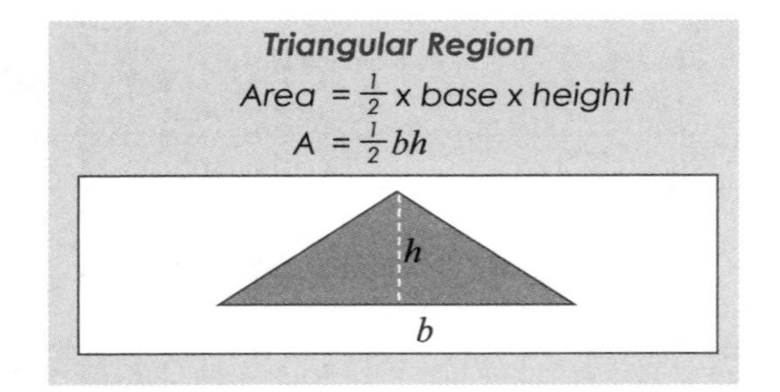

Circular Region

Area = Pi x radius x radius

A = πr^2

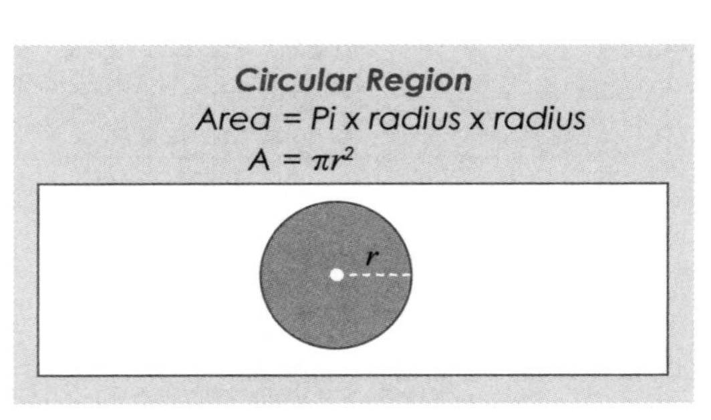

Elliptical (Oval) Region

Area = Pi x a x b

A = πab

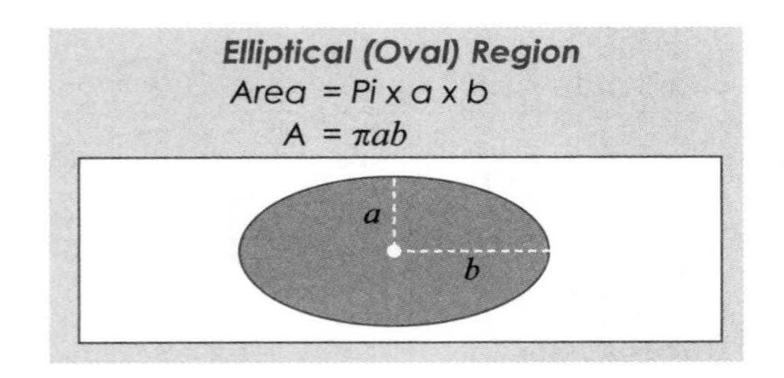

Rules for Finding Surface Area

Sphere

$A = 4\pi r^2$

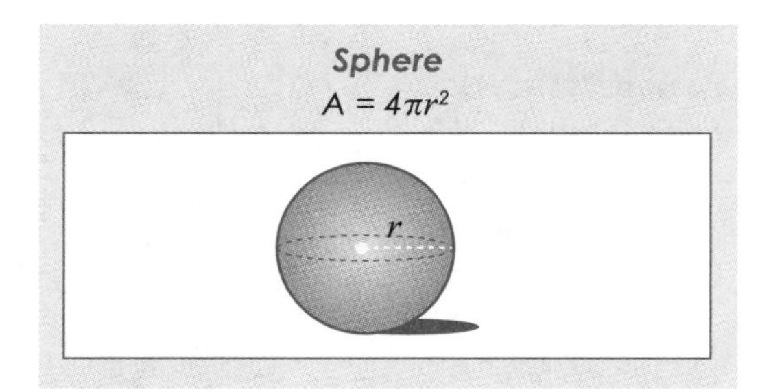

Cube

Area = 6 x Area of one face

$A = 6s^2$

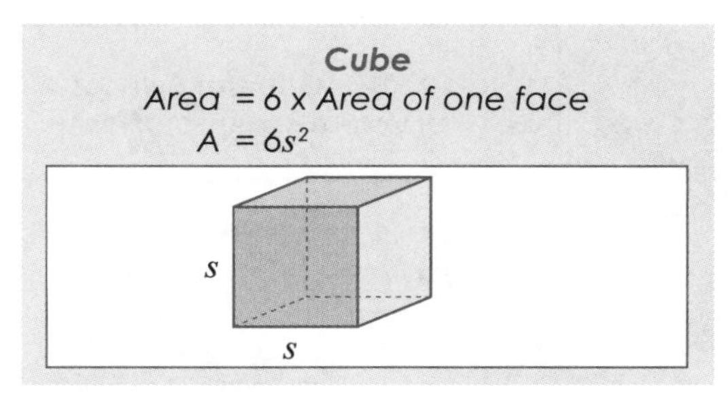

Cylinder

Area = (2 x area of base) + Area of curved surface

= (2 x area of base) + (circumference of circular region x height)

$A = 2\pi r^2 + 2\pi rh$

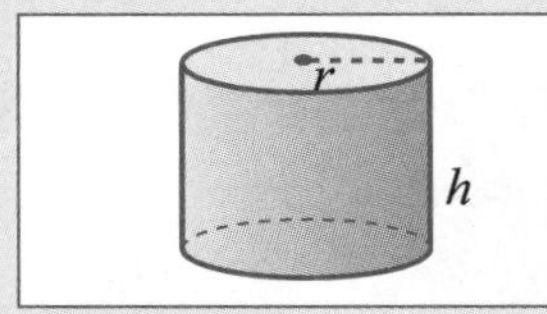

Rectangular Prism

Area = 2lw + 2lh + 2wh

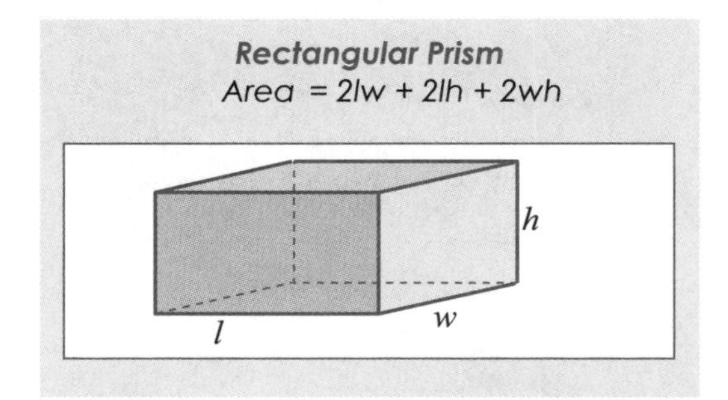

Cone

Area = Area of base + Area of curved surface

$A = \pi r^2 + \pi rs$

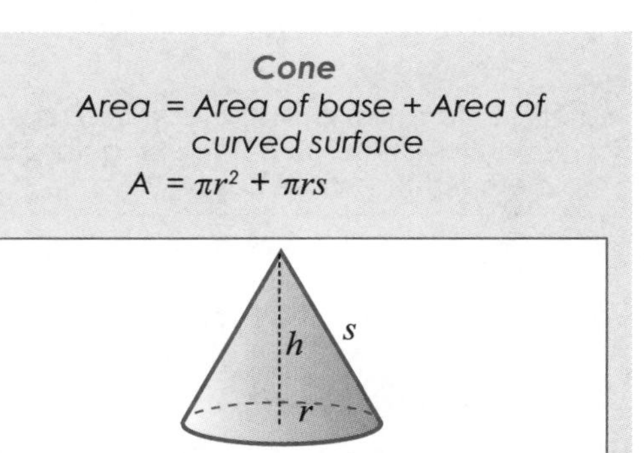

Hemisphere

Area $= \frac{1}{2}$ x (Area of sphere) + Area of circular region

$= \frac{1}{2}(4\pi r^2) + \pi r^2$

$= 2\pi r^2 + \pi r^2$

$= 3\pi r^2$

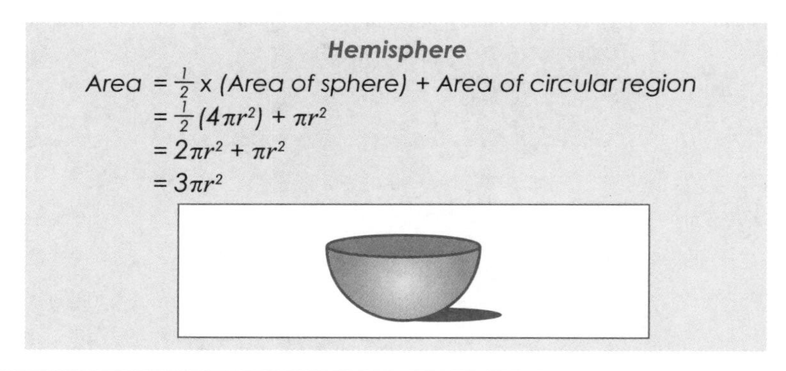

Rules for Finding Volume of Prisms

General Rule for All Prisms and Cylinders
Volume = Area of base x height of prism or cylinder

Rectangular Prism

Volume = *length* x *width* x *height*
= $l \times w \times h$
V = lwh

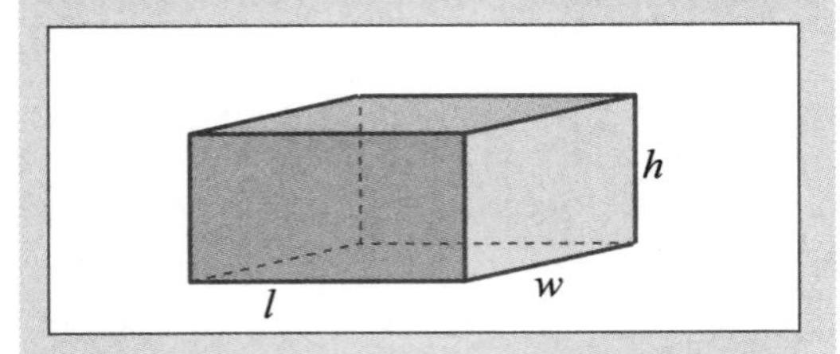

Cube

Volume = *length* x *width* x *height*
= $s \times s \times s$
V = s^3

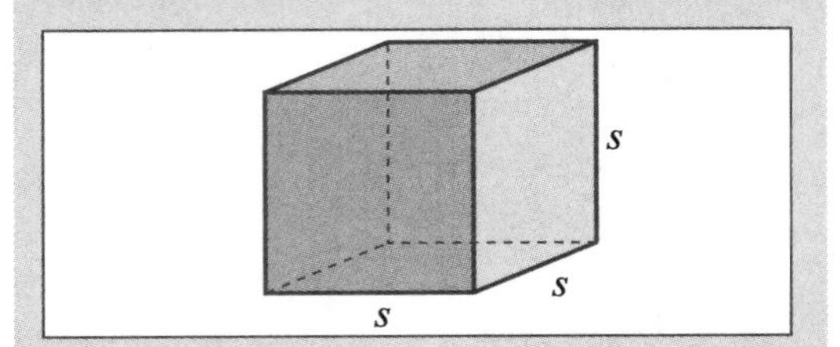

Triangular Prism

Volume = *Area of base x height of prism*
= $\frac{1}{2}$ x *base of triangle x height of triangle x height of prism*
= $(\frac{1}{2} \times b \times h) \times l$
V = $\frac{1}{2} bhl$

Note: For triangular prisms, two heights are involved, the height (h) of the triangular base and the height (l) of the prism.

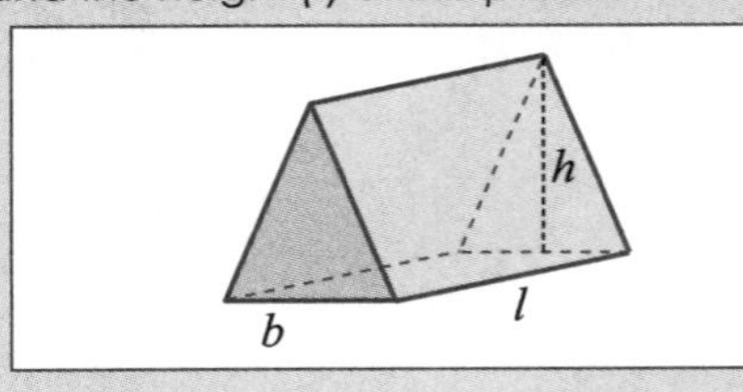

Trapezoidal Prism

Volume = *Area of base x height of prism*
= $\frac{1}{2}$ x *sum of parallel side lengths of prism trapezoid x height of trapezoid x height of prism*
= $[\frac{1}{2}(a + b) \times h] \times l$
V = $\frac{1}{2}(a + b)hl$

Note: For trapezoidal prisms, two heights are involved, the height (h) of the trapezoidal base and the height (l) of the prism.

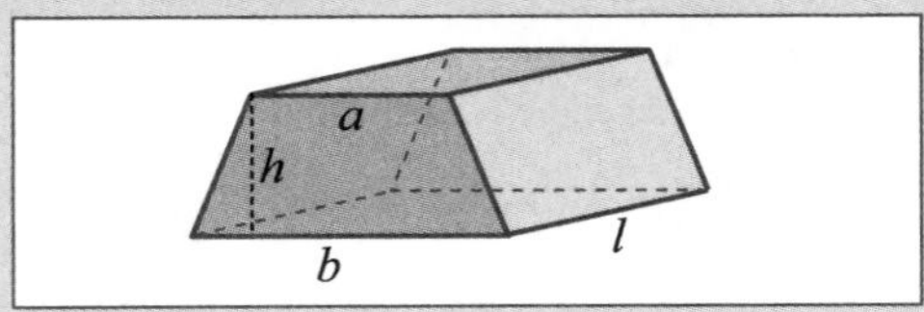

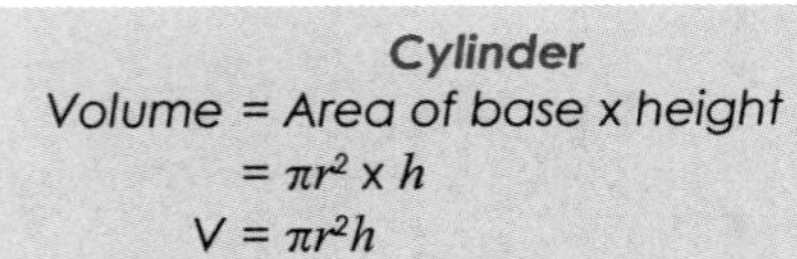

Cylinder

Volume = *Area of base x height*
= $\pi r^2 \times h$
V = $\pi r^2 h$

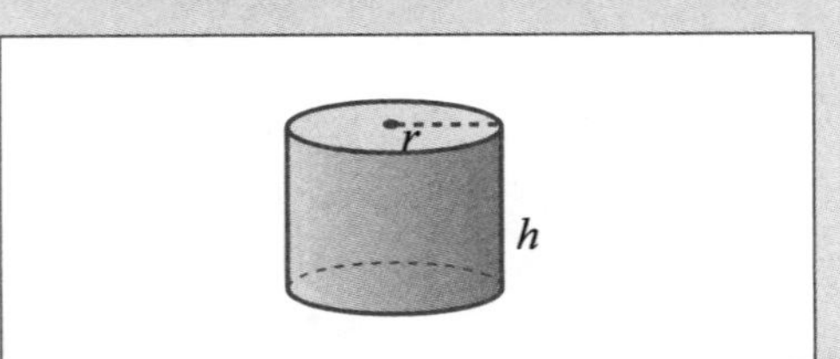

Rules for Finding Volume of Pyramids & Spheres

General Rule for All Pyramids and Cones

Volume = $\frac{1}{3}$ x *area of base* x *height of a pyramid or cone*

Rectangular Pyramid

Volume = $\frac{1}{3}$ x area of base x height

= $\frac{1}{3} \times l \times w \times h$

V = $\frac{1}{3}lwh$

Square Pyramid

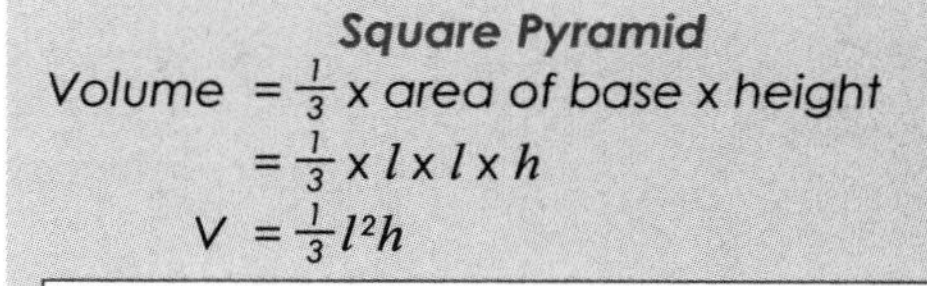

Volume = $\frac{1}{3}$ x area of base x height

= $\frac{1}{3} \times l \times l \times h$

V = $\frac{1}{3}l^2h$

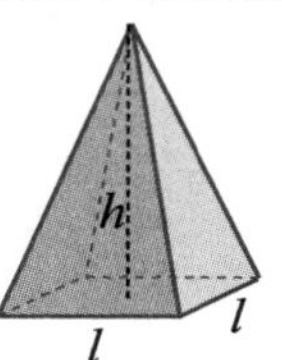

Triangular Pyramid

Volume = $\frac{1}{3}$ x area of base x height

V = $\frac{1}{3} \times \frac{1}{2} l \times w \times h$

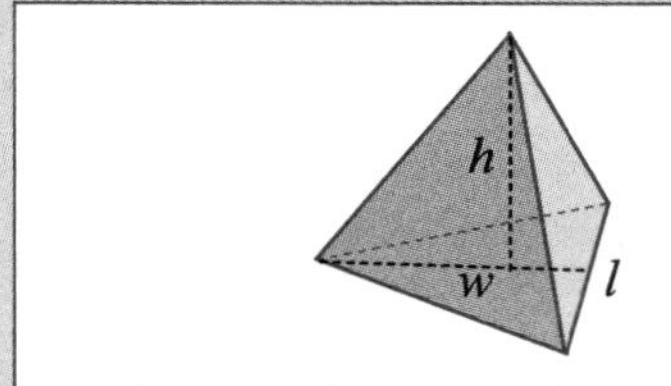

Cone

Volume = $\frac{1}{3}$ x area of base x height

= $\frac{1}{3} \times \pi r^2 \times h$

V = $\frac{1}{3}\pi r^2 h$

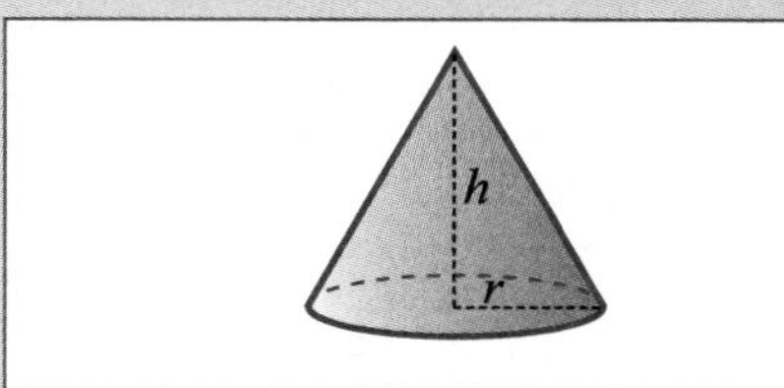

Sphere

Volume = $\frac{4}{3}\pi r^3$

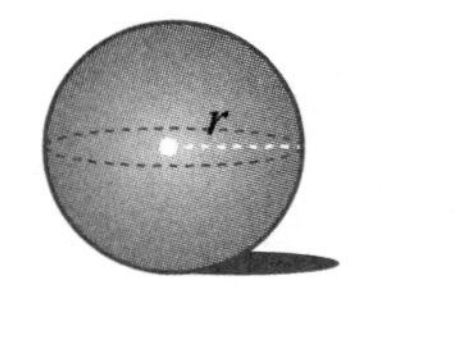

Common Math Symbols

Symbol	Meaning
$=$	equals, is equal to, is the same as, is
$\neq$	is not equal to
$\approx$	is approximately equal to
$>$	is greater than
$\geq$	is greater than or equal to
$<$	is less than
$\leq$	is less than or equal to
$\cong$	is congruent to
$a : b$	ratio of a to b or $^{a}/_{b}$
π	pi, approx 3.14 or $^{22}/_{7}$
%	percent
$\sqrt{}$	the square root of
$\sqrt[3]{}$	the cube root of
$0.\overline{3}$	0.33333 ... ; or 0.3 repeating
!	factorial; 3! = 3 x 2 x 1
∞	infinity
$^{\circ}$	degree
Δ	triangle; ΔABC means triangle with vertices A, B and C
$\angle$	angle; $\angle$ABC means an angle with vertex B
m$\angle$A	measure of angle A
$\perp$	is perpendicular to
⦜	a right angle
m$\overline{AB}$	measure (length) of a line segment

5-19